PROJECT SUNSHINE

PROJECT SUNSHINE

How science can use the sun to fuel and feed the world

STEVE MCKEVITT & TONY RYAN

ICON

Published in the UK in 2013 by
Icon Books Ltd, Omnibus Business Centre,
39–41 North Road, London N7 9DP
email: info@iconbooks.net
www.iconbooks.net

Sold in the UK, Europe and Asia
by Faber & Faber Ltd, Bloomsbury House,
74–77 Great Russell Street,
London WC1B 3DA or their agents

Distributed in the UK, Europe and Asia
by TBS Ltd, TBS Distribution Centre, Colchester Road,
Frating Green, Colchester CO7 7DW

Distributed in India by
Penguin Books India,
11 Community Centre, Panchsheel Park,
New Delhi 110017

Distributed in South Africa by
Book Promotions, Office B4, The District,
41 Sir Lowry Road, Woodstock 7925

Distributed in Australia and New Zealand by
Allen & Unwin Pty Ltd,
PO Box 8500, 83 Alexander Street,
Crows Nest, NSW 2065

Distributed in Canada by
Penguin Books Canada,
90 Eglinton Avenue East, Suite 700,
Toronto, Ontario M4P 2Y3

ISBN: 978-184831-513-6

Typeset in ITC Galliard by Marie Doherty
Printed and bound in the UK by Clays Ltd, St Ives plc

Contents

About the authors

Steve McKevitt is the author of *Everything Now, Why the World is Full of Useless Things* and *City Slackers*. His writing has appeared in newspapers around the world from *The Guardian* to the Kenya *Daily Nation*. An expert in communications and consumerism, over a 25-year career his clients have included Nike, Coca-Cola, Deutsche Bank, Sony PlayStation, Harvey Nichols, Motorola, Universal, Virgin, BT and Atari. Steve also works as an advisor to national and regional UK government on employment, skills, business innovation and international trade.

Tony Ryan is a chemist with a specialism in polymer science. He is a Pro Vice Chancellor at the University of Sheffield, where he leads the Faculty of Science. He was previously the ICI Professor of Physical Chemistry and gave the Royal Institution Christmas Lectures in 2002. Tony's research is wide-ranging, encompassing synthesis of polymers from renewable feedstocks, the design of drug delivery vehicles, scaffolds for tissue engineering, nanotechnology in home and personal care, the fundamentals of crystallisation, and the self-assembly of block copolymers. He has served on numerous advisory boards and councils and was until recently Chair of the Science Board of the Science and Technology Facilities Council. In 2006 he was made an Officer of the British Empire for Services to Science.

*To our wives Fiona McKevitt and Angela Ryan
and our mothers Thelma McKevitt and Margaret Ryan*

Acknowledgements

Project Sunshine arose philosophically from a dawning realisation in both of us. Science and engineering would need to provide the tools for a political and economic system that could deliver well-being to a burgeoning population: enough food to thrive and enough energy to prosper, at the same time as ameliorating the damage we have done to the biosphere that's just a thin veneer on the planet. This realisation occurred to each of us separately, but essentially concurrently.

Project Sunshine happened in practice because there was a reorganisation at The University of Sheffield that allowed an integrated view of the capabilities and potential for co-ordinated research. And this book was the outcome of Steve's company Golden being commissioned to make a short film featuring Tony that espoused that vision for the future.

Many University of Sheffield researchers have given generously of their time and expertise: Peter Horton, Duncan Cameron, Julie Scholes, Paul Quick, Colin Osborne, Andy Fleming, Jonathan Leake, Ian Woodward, David Beerling, Terry Burke, Rob Freckleton, Julie Gray, Neil Hunter, Jim Gilmour, Will Zimmerman, Chris Jones, Róbert von Fáy-Siebenbürgen, Sean Quegan, Mark Geoghegan, Richard Jones, Alastair Buckley, David Lidzey, Alan Dunbar, Ahmed Iraqi, Julia Weinstein, Peter Hall, Peter Jackson, Mike Braddick and Steve Banwart. In the Faculty of Science, Angela Simonite, Chris Smith, Sarah Want, Shelagh Cowley and Terry Croft made things happen. And from outside Jim Barber, David Phillips, Athene Donald, Tony Ryan (senior), Helen Storey and Nate Lewis helped in more ways than they could have known. Phil Waywell deserves a special mention for his many contributions and the reading of early drafts.

Writing this book together has been a revelation, and we are grateful for the support and patience of our families. We were looked after by Steph Ebdon of the Marsh Agency and very lucky to have Duncan Heath as our editor at Icon.

Preface: Tick Tock

'Until about 1800, our species had no safety margin
and lived, like other animals up to the limit of the
food supply, ebbing and flowing in population.'
Jeremy Grantham, investment specialist

For almost all of our species' 200,000-year-long history,
man's relationship with the Earth was no different to that
of any other creature. Humans have a unique ability to create and communicate, which gave our early ancestors significantly more control over the environment and allowed
them to build sophisticated societies. Yet, in respect of what
they took from the world and how they interacted with it,
they were very much like the birds, bees and chimpanzees: all
their energy was provided directly by the sun. Sunlight captured by photosynthesis was converted into food and fuel.
They ate roots, grains, fruit and cereals to provide them with
energy (or to feed to animals, which, in turn, they also ate).
They burned wood to keep themselves warm and vegetable
oil or animal fat to provide night-time light. It was a successful strategy for survival and over tens of thousands of years
the human population spread across six continents. However,
locked in to this natural solar cycle, there was a limit to how
many people their lifestyle could support, and the total number of inhabitants fluctuated between just 300 million and
500 million, with this variance in the sustainable maximum
accounted for by the usual suspects: war, famine, plague and
pestilence.

Then, 350 years ago, everything changed. We began to
supplement our energy needs with coal and oil; the same
captured sunshine, but this time millions of years old, preserved in fossilised form deep underground. In less than two
centuries the human population had exploded, doubling in
size to 1 billion people. It has continued to grow ever since,
but the rate of change has increased significantly. It took

100,000 years to reach the first billion people: today we are adding further billions at a rate of one every twelve years. The result is a huge squeeze on all natural resources. Over the next two decades we will witness a 50 per cent increase in demand for energy, food and water.

We are now living through our fourth century of exponential population growth, but the solutions of the past that have allowed us to cope – burning more oil, gas and coal; expanding the amount of land under cultivation – simply won't work any more. Our stocks of fossilised sunlight are dwindling, we are running out of fresh water and of places that we can appropriate for farming. Looking to the future, the key question is not simply, How many people are there going to be?, but rather, How are they all going to live?

In 2012, the Royal Society sought to address this issue in a publication called *People and the Planet*. This report's frank conclusion was that in the developed and emerging economies, consumption has reached unsustainable levels and must be reduced immediately. It claims that the increase in population will 'entail scaling back or radical transformation of damaging material consumption and emissions and the adoption of sustainable technologies. This change is critical to ensuring a sustainable future for all.'

We have been here many times before. The whole of human history is essentially the story of population growth and increasing competition for resources. Since ancient times wars have been fought over land, water, food, fuel, metals and other resources, while many civilisations have also faced destruction as a result of disease, famine and shortages. Where we have overcome these challenges in the past it is invariably ingenuity and innovation that have provided the solution. And we will require those qualities in abundance, because we have never had to tackle problems of this magnitude before, nor on a global scale.

What it will take to sustain a world of 9 billion people is the subject of this book. Achieving this goal is possible and realistic, but it will not be easy. It will not happen by accident and it will bring changes for all of us both in the way that we live and in what we consume. It is without doubt the biggest challenge of our age. This statement is not intended to belittle the impact of climate change, but the repercussions of global warming will play out along a timescale of decades; we have some time to adapt and respond to the consequences. In contrast we are already living with the results of explosive population growth: rising fuel and food prices, wars, immigration, famine, energy shortages and economic uncertainty. These issues affect us all today and all are a direct consequence of adding 1 million people to the population of the developing world every five days.

Big problems require radical solutions, but there are genuine grounds for optimism. Fundamentally, we need to reconnect the global economy with the sun and live within our means, just as we did in the past. Capturing a single hour of the sunlight that reaches the Earth – a tiny fraction of our star's output – would meet our global energy needs for a whole year. Harnessing the power of the sun will allow us to meet the increasing food and energy needs of the world's population in the context of an uncertain climate and global environment change.

To discover how we can achieve this requires a change in the way that scientists think and work, crossing the traditional boundaries in both the pure and applied sciences and engaging in collaborative research and innovation. Thankfully this is already happening, and in the later sections of the book we explore how the fruits of this activity are already being used in this endeavour: mathematicians are unlocking the secrets of how the sun actually generates its energy so that sunshine can be captured to fuel and feed the world; physicists are developing photovoltaic devices to convert this sunshine into electricity that we can use in our homes and businesses; chemists are investigating how plants and algae could be used to produce alternatives to fossil fuel in the form of biogas and biodiesel; biologists are learning to improve the efficiency of photosynthesis itself to achieve a better yield from crops that will require less fertiliser, water and pesticide; and engineers from all disciplines are working out how to put the results of all this research into practice. Success will require a concerted effort across the spectrum, not just from academics but from policymakers and private enterprise as well, to bring to market as quickly as possible the products that can make a real difference.

Our message is profound and optimistic, but not profoundly optimistic. Success entails that a world containing 9 billion people will have to be very different to the one we are living in today. Different, but not necessarily worse. Sustainable routes to food and energy security can be found, but time is of the essence. The clock is ticking.

1   Seven Billion and Counting

> 'The constant effort towards population, which is
> found even in the most vicious societies, increases
> the number of people before the means of
> subsistence are increased.'
> *Thomas Malthus, essayist*

On 30 October 2011 the world welcomed its seven-billionth
citizen: Danica May Camacho, a Filipina, born in the early
hours of the morning at the Dr Jose Fabella Memorial
Hospital in Manila. She was chosen by the United Nations
Population Fund (UNFPA) to officially mark this milestone and draw attention to the economic, social and practical challenges of managing the world's rapidly growing
population.

The fact that these challenges require such a grandiose
PR stunt to make them newsworthy at all is testament to the
fact that we find them so easy to ignore. For most of us, the
Official Day of Seven Billion was a story to be forgotten as
soon as the agenda moved on to something else. Danica May
Camacho herself, briefly the most famous baby in the world,
is likely to live out the rest of her life in the obscurity endured
by twelve-year-old Adnan Nevic of Bosnia Herzegovina and
Matej Gaspar, a 24-year-old Croat, respectively the world's
six- and five-billionth inhabitants. It's easy to understand
our indifference. On first inspection, there seems very little
truly new to say on the subject. The quote from Thomas
Malthus that opens this chapter sounds like it could have
been uttered last week rather than 1798. This is because
the fundamental issue remains the same: too many people/
not enough resources. One could be forgiven for thinking
that little else has changed since Malthus penned his *Essay
on the Principle of Population*; the same dire warnings about
famine and drought, the same apocalyptic forecast of global

wars that will bring about the collapse of civilisation, and the same list of unspeakably miserable consequences for us all: none of which has come to pass. However, this scepticism is misplaced.

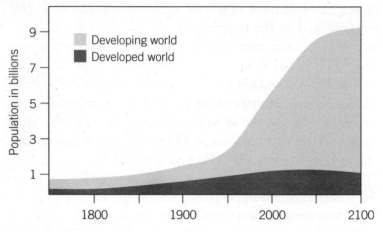

Fig. 1: World population growth since 1750. (Source: UNFPA)

It's true that all our lives have played out through a period of explosive population growth, but that doesn't stop it being extraordinary. Indeed, this is also the *only* period of explosive population growth in human history. We have taken just 84 years to go from 2 billion to 7 billion earthly inhabitants and, unless we take some decisions about how everyone is going to have to live, we will soon reach a point where the global population becomes unsustainable. While forecasting is a notoriously contentious and difficult discipline, among those who are looking to the future there is a consensus that global population will continue increasing until the middle of the century, at which point it will peak and plateau at somewhere between 9 and 10 billion. From less than 2 billion to 9 billion people in little more than a lifetime; the blink of an eye when set against the 200,000 years that our species, *Homo sapiens*, has been on the planet.

It's tempting to believe that a simple presentation of

the facts will be enough to shake us from our complacency. Certainly, one would assume that was the rationale of the UNFPA when it conceived the idea of Citizen Seven Billion, but unfortunately there's more to our indifference than this. It's not just that over-familiarity makes this story easy to ignore, it's that most of us choose actively to ignore it. We feel reassured by the trappings of our advanced society with its central heating, running water, supermarkets, ready meals and fuel-injected cars; at a comfortable remove from the sources of food and energy. Yet despite our apparent sophistication, in evolutionary terms we have barely set foot out of the forest. We are the same nervous, skittish creatures that were once hunted mercilessly by leopards, wolves and cave bears, with the same reactions to fear and danger.

While we have the intellectual capacity to think about the future and ponder, 'What might happen if …?', we are much more focused on the present; driven by today's needs rather than tomorrow's consequences. As a result we have evolved to be remarkably good at ignoring 'What might happen if …?', especially if we suspect that thinking about it might prove terrifying. The smoker enjoying the first cigarette of the day; the commuter racing down the motorway at 85mph and the student choosing an evening out over a night of revision, are all aware of the possible consequences of their actions at a nebulous point down the line (lung cancer, a car crash, examination failure) but that only makes them easier to disregard. We treat these as things that will happen to other smokers, other drivers and other revellers, not to us. In these cases and many others, this wilful ignorance is bliss. We behave in exactly the same manner when confronted by less personal or unspecific dangers; it's really just a question of the scale of our denial.

As far as threats with terrifying, immeasurable consequences go, global warming takes some beating. In 2007, Al Gore visited the University of Sheffield to host a conference on climate change. We were fortunate enough to receive an invitation to attend. Whatever your political views, there's no denying that Al Gore is a very capable and engaging public speaker. Over the course of 90 minutes, the former US Vice President performed a live version of his Oscar-winning documentary, *An Inconvenient Truth*, during which he clearly laid out all the evidence for human influence on global climate change and explained its consequences. His compelling argument and powerful delivery certainly made for a fascinating lecture, but also for one of the most dispiriting things we have ever seen.

Gore had two stated objectives for *An Inconvenient Truth*. He wanted to leave audiences believing that global warming is the biggest issue, but he also wanted to persuade them to change their behaviour by making them believe that doing so could help to reverse its effects. To that end, he concludes the film with the following call for action:

> Each one of us is a cause of global warming, but each one of us can make choices to change that with the things we buy, the electricity we use, the cars we drive; we can make choices to bring our individual carbon emissions to zero. The solutions are in our hands, we just have to have the determination to make it happen. We have everything that we need to reduce carbon emissions, everything but political will.

While there's little doubt that he achieved his first objective, he has been much less successful in changing our behaviour. For a week or so following the live lecture, we felt deeply depressed about not only the future, but the futility of our own efforts to reverse the effects of global warming (switching to low-energy light-bulbs, unplugging electrical appliances when not in use, driving at 60mph instead of 70mph – that sort of thing) in the face of the two coal-burning power stations that were being opened in China each week. Within a fortnight, however, we were back to our usual chipper selves, thanks not to thinking of creative solutions to reduce our own carbon footprint, but simply to not thinking about it at all.

Of course ignoring the problem isn't going to make it go away, but then neither will worrying about it. The most frustrating aspect of all the challenges we face – not just climate change and population growth, but food and energy sustainability as well – is that we already have all the science and technological solutions to avert disaster; what we lack is the social and political will to implement them. Moreover, if a combination of complacency, fear, and wilful ignorance makes it difficult for us to motivate ourselves and our politicians towards effective action, conversely it provides a fecund opportunity for those wishing to persuade us, however disingenuously, that everything is going to be all right.

One doesn't have to look very hard to find climate change 'sceptics' who focus on minor flaws in the science that 'undermine the entire argument'. Just because there are naysayers doesn't mean there needs to be a debate. For example, despite all evidence to the contrary, if you type 'smoking doesn't cause cancer' into Google, your search will yield 143,000 results, all purporting to prove that it doesn't.

It is noteworthy that almost all the scepticism about climate change comes from the conservative right. Surely if the data was so equivocal, one would expect dissenters across the political spectrum? Yet, for whatever reason, this is not the case. These outliers often find a platform for their views that is vastly disproportionate to their credibility, because what they claim is much closer to what most people want to believe is true: that the status quo will be maintained. It's certainly what we'd like to believe is true, but unfortunately, in the face of all the evidence, that is impossible to do.

Our fear of change prevents the adoption of potentially life-saving technologies such as nuclear power, concentrated solar energy and genetically modified plants and animals. We can argue about how much oil is left until finally someone is right and there isn't enough; and we can argue about whether to genetically modify crops until there's nothing left to eat. Alternatively, we can act now.

We will certainly reach a point from which it will be

impossible to recover, but we are not there yet. It really doesn't need to end unhappily. And now for the good news – there is something that we can do about it.

※

Science and technological innovation have driven global prosperity. Since the Enlightenment of the 18th century they have proved consistently capable of meeting the ever-increasing demand for energy and food. In the past 40 years alone, the amount of land used for agriculture has increased by only 8 per cent, while food production has doubled. This success is almost entirely due to chemical and biological breakthroughs and innovations: providing more effective pesticides and fertilisers; improving crop and meat yields through breeding programmes. We must ensure that the fruits of this process of invention are sustainable.

The challenges we face in the next 40 years are complex and difficult, but they are not insurmountable. A study published in January 2011 by the UK's Institution of Mechanical Engineers suggested there are no scientific breakthroughs required to manage a global population of over 9 billion people:

> There is no need to delay action while waiting for the next greatest technical discovery or breakthrough idea on population control ... [There are] no insurmountable technical issues in meeting the basic needs of nine billion people ... sustainable engineering solutions largely exist.

There are key areas that we need to focus on to ensure food security. The huge improvements in crop yields have to continue, but they are eminently deliverable. We must make less profligate use of our fresh water supplies; and develop genetic solutions to crop protection and rely much less on chemical fertilisers and pesticides. We need to develop a system of agriculture that is holistic, part of the richer ecosystem rather than the wilfully imposed cereal monoculture we have today that operates outside it. Livestock and marine food production can continue only within the context of sustainability. We can also make a big difference by choosing to live less wasteful lifestyles: currently in the developed nations, over a third of all food that is harvested is simply thrown away.

Providing energy security is rather more complex, particularly because almost all of the fuels we use presently are the major contributors to climate change. We have around two decades to de-carbonise electricity generation, which will require significant investment in emerging technologies and processes. Achieving holistic solutions will require scientists from different disciplines working together across traditional boundaries.

Project Sunshine is the story of how we are going to provide sustainable food and energy for a global population of 9 billion people. The answers lie in a range of ongoing research across many disciplines: from solar physics, photovoltaics and photosynthesis to plant physiology, biochemistry and ecology. This research is typically disparate, very detailed and difficult to digest. We will provide a real solution only by pulling it all together and putting it into context. This book aims to do exactly that.

All of our energy comes from the sun. Fundamentally. We need to understand how the sun works, how it provides us with that energy, and learn how to use some of that energy to power everything that happens on earth in real time, rather than relying on ancient sunshine stored in coal, oil and gas. In part, this is about unlocking the mysteries of electrons, molecules and genetics, but we also need to take a much grander view, to understand how carbon, nitrogen and phosphorous are traded on a global level. And so our quest will begin and end with mathematics and theoretical physics. By bringing the many pieces of research together and synthesising them, we can get a true picture of how we are going to live – and going to have to live – in the future. There's no point in worrying. The future is going to be very different, but that doesn't mean it's going to be scary, or even worse. There's no reason to fear that you'll be living out a real-life version of *The Road* any time soon.

To show how we can safeguard the future we need to understand how our world became the place it is today – chemically, geologically, ecologically, climatically and economically. We need to understand where all our food and energy comes from, to help us decide what we need to live and what we can live without. Among all the animals, we have the unique ability to change the environment for the benefit of the species, but if humankind is to survive and prosper we will have to do this more effectively and sustainably. We will have to start living within our means again, rather than beyond them: to go forwards we need to go back to a solar economy.

2   Weathering a Perfect Storm

'There are dramatic problems out there, particularly with water and food, but energy also, and they are all intimately connected. You can't think about dealing with one without considering the others. We must deal with all of these together.'
John Beddington, chief science advisor to the UK government

In 2009, the UK Office for Science published a paper called *Food, Energy, Water and the Climate: A Perfect Storm of Global Events?* Written by John Beddington, the UK government's chief science advisor, *A Perfect Storm* is a harrowing document. At least it is if you take it at face value, which is exactly what the media did. In summary, the report highlights the fact that the world's projected population growth by 2030 will lead, together with the incumbent economic and environmental factors, to a 30 per cent increase in demand for water and a 50 per cent increase in demand for food and energy. The press went to great lengths to ensure that the scale of these challenges was not understated. Good news sells few papers and in that regard *A Perfect Storm* made for excellent copy. Yet there's another way of looking at the information contained in the report. Beddington's erudite analysis of the challenges is certainly not bedtime reading for those of a nervous disposition, but the report's real success, and one rarely noted, is that it highlights everything that we need to do: as the starting point for a strategy to address these challenges, it could not be better.

A Perfect Storm describes not the end of the world, but a starting point for its salvation. It's definitely not the best place to start from, but the most important thing is that we do know where to start. The implications of population growth will not prove to be as easy to ignore for much longer.

New inhabitants are being added at the rate of 6 million each month (the equivalent to a city the size of Rio de Janeiro) and they will not be spread evenly. In the developed world outside the USA low birth rates mean that indigenous populations are in decline in many countries. By 2020, there will be more people over the age of 60 than under 20 in many European states. Conceived in the 1940s, the UK's welfare state, which served as a model for many other countries, was designed to cater for citizens who worked for 50 years, paying national insurance while they did, then spent a few brief years in retirement drawing it out before cost-effectively and expediently passing away.

Without reform, this shift in demographic is going to put national health and welfare systems under enormous pressure. Today most people can reasonably expect to live for another 20 or even 30 years after retiring. This means that over the next 30 years, there will be more and more people leaving economic productivity and entering retirement, but conversely fewer and fewer people of working age to look after them. There will also be fewer people to pay for national insurance to meet the increased demand for healthcare and welfare.

The key points here are that we already know this – it's not some nasty surprise waiting to pounce – and we have 30 years to sort it out. To do that we need to plan for this future today and completely rethink the way that welfare is funded. Yes, there will be more responsibility on individuals to plan and pay for their retirement and healthcare, and services we take for granted today will become much more expensive, but we have the luxury of being able to decide now what we want to deal with this future problem and how we want to pay for it. We don't have to fire-fight or come up with policies on the hoof.

While Europe deals with its burgeoning pensioners, many parts of the developing world are dealing with the opposite problem: explosive population growth that is outpacing economic growth. In Africa the continent's population is set to double from 1 billion to 2 billion by 2030, by which point half of its inhabitants will be under the age of 20. Similar rapid expansions are being experienced across much of the developing world, provoking other transformational changes, most notably urbanisation, as people migrate to cities from rural areas in search of work. Half the world is already living in cities, but this will increase to 60 per cent by 2030. There will be at least 29 'mega cities' with more than 10 million inhabitants by 2025. That is ten more than there are today. All of these additional people will require food, water, shelter, energy and a host of other services. The inevitable competition for land that this will create is foreshadowed today by increasing purchases of real estate in developing nations by some countries with hot and dry climates and limited water supplies – notably Egypt, Saudi Arabia and China – and also by multinational corporations.

Our challenge to ensure food security for a global population of 9 billion in a manner that is equitable, healthy and sustainable is simple to understand: we must grow much more food on the same land, using less water, fertiliser and pesticides than ever before. Achieving this in the face of rapidly dwindling natural resources will be no walk in the park, but it is possible. Science and technology will make the most significant contribution, providing practical solutions across the board from engineering to biotechnology, but success will also demand behavioural changes of us all. In future, we will need to manage our lives and societies much more efficiently than we do at the moment.

2008 saw the sudden end of a 20-year economic Golden Age for almost all of the world's leading economies. It's easy to be critical with hindsight, but at the time most governments, financiers and economists were of the view that perpetual and sustainable affluence was a realistic objective – 'the end of boom and bust' no less. The Credit Crunch brutally dispelled our beguilement by the arcane chicanery of banks and other financial institutions, bringing the blunt realities of commodities markets into sharp focus. Wheat and maize prices rocketed before finally settling down at three times their 2005 levels, thereby marking the end of two decades of low-cost food for consumers. Although prices are less volatile today, cereal stocks remain stubbornly at a 40-year low, which, together with the increasing demand for food, energy and water from the emerging economies, will continue to put pressure on food prices for the foreseeable future. Increasing the yield of cereal crops using existing, proven technologies is both practical and realisable in the short term. This should be the goal for every one of the world's agricultural ministries.

Not every nation was plunged into recession in 2008. Those states operating more traditional economic models – based on production, manufacturing and government intervention, rather than leveraged borrowing, notional property values and unfettered money markets – continued to perform well, particularly those with large populations and plentiful natural resources. Brazil, Russia, India and China are known collectively as the BRIC countries. These states have all embraced global capitalism and adapted their political systems to facilitate rapid growth. It is expected that they will become the dominant suppliers of manufactured goods and services over the coming decades, with Russia and Brazil also becoming the dominant suppliers of raw materials. The BRICs' burgeoning prosperity is a further, powerful driving force behind the demand for energy. Since 1900 real income has grown by a factor of 25, and primary energy consumption by a factor of 22.5. Natural resources are in decline and competition for what's left is going to increase as 1 billion super-consumers of the OECD are joined by a further three billion from the BRICs.

Economic success, within the BRICs at least, will lead directly to an increase in prosperity, lifting tens if not hundreds of millions of people out of poverty. But this positive outcome will only add to our list of challenges. When wages rise in developing and middle-income countries, we find that people consume more meat and dairy products, which in turn causes rapid growth in demand for agricultural commodities to feed the extra livestock. The continued pressure on cereal stocks is due in part to the rising consumption of meat and dairy, especially in China and Brazil. There are no signs that cereal prices will flatten out any time soon, indeed it's reasonable to expect that they will continue to increase, as incomes grow in India and sub-Saharan Africa; places where per capita meat consumption today is low. The UN Food and Agriculture Organisation (FAO) projects that farms will be required to produce around 40 per cent more food to meet the demand in 2030 than they did in 2008. Yet even this startling estimate – the equivalent of an annual increase in productivity of 1.5 per cent – is not the whole story. Meat demand will double by 2050 and all those additional animals will also need feeding. Each of these production targets has further implications for supplies of land, water and most importantly of all, for supplies of energy.

This pressure on agriculture to produce more with less will make the biggest contribution to a 45 per cent increase in demand for energy between 2006 and 2030. Notwithstanding the real situation regarding how much oil is left in the ground, mitigating climate change means that an alternative to fossil fuel will be required to make up the (significant) shortfall. Biofuels can be used for transportation, while biomass can be burned to produce heat or electricity. This will, however, provide even greater competition for land, water, food and energy. Again, the majority of this demand for energy is going to come from within the BRICs and notably from India and China, which between them contain approximately half of the world's people.

Like food, water demand is a function of population, incomes, diets and the requirements of irrigated agriculture, but also of industrialisation. Heavy industry – like the kind powering growth in the BRICs – requires lots of water. Agriculture will find itself increasingly competing for water and land not just with commerce, but with the cities it's being asked to feed. Mid-range estimates suggest that the demand for fresh water by agriculture alone will be 30 per cent higher in 2030, while the total global demand could be as much as 60 per cent higher by 2025. Shortly, we will find water being treated like any other commodity and subjected to the same market forces. The notion of a free natural resource will be consigned to the history books and bottled water will become a fact of life rather than a fad or affectation. Today 1.2 billion people are already living in areas affected by water scarcity; this figure will increase significantly in future. There are already early signs of things to come. Water conflict occurs between two or more neighbouring countries that share a trans-boundary water source, such as a river, artesian basin or lake. In the case of Kazakhstan, Uzbekistan, Turkmenistan, Tajikistan and Kyrgyzstan, the dispute is over access to the Aral Sea. With no satisfactory diplomatic solution on the horizon, relations between the five nations are increasingly hostile.

Our entire industrial and agricultural system relies upon a constant supply of oil. We use oil and liquid fuels at a rate of 89 million barrels per day (mb/d); demand in 2030 will be at least 25 per cent higher. There isn't an infinite supply of oil, and even if we are finding more creative ways to extract what's left, one day we will run out. It's impossible to know for sure how much oil remains, but just because we may have underestimated the quantity in the past, that doesn't mean we're underestimating how much is there now. Unfortunately we won't know for certain until we reach a point where we can't extract enough to satisfy demand. 'Peak oil' is the term used to describe this point: when the maximum rate of global petroleum extraction is reached, after which the rate of production enters into terminal decline. The Day of Peak Oil is even more difficult to estimate than the Day of Seven Billion – we can 'see' the people, but the world's reserves of oil remain hidden from us. The idea of peak oil is based on observations of production from existing

oil wells and fields combined with estimates about the likelihood and size of undiscovered reserves. We may have already
hit peak oil. The International Energy Agency believes that
2006 was the peak year of production for conventional crude
oil, and even the most optimistic estimate forecasts that production will decline after 2020.

An uneasy balance also characterises the oil market. Most
or our oil comes from the more politically turbulent areas
of the world, while demand fluctuates dramatically. During
recessions, industry and consumers use less oil so the market
softens; likewise the price increases during periods of growth
when demand is greater. Other factors such as the weather
can also dramatically affect the price. January 2012 was a
typical month, with tensions surrounding Iran counteract-
ing a weaker economic outlook. In Europe the late onset of
winter weather pushed prices for Brent Crude to six-month
highs in early February, trading at \$117.50/barrel. In con-
trast, slower than expected demand from industry led to rising stocks at some storage depots, pressuring the price of
West Texas Indeterminate – a lighter oil variant than Brent
– down to \$99.50/barrel.

The key point is not just that the supplies of oil are
decreasing, but that the global demand for whatever is left is
increasing. Global oil demand rose to 89.9 mb/d in 2012,
a rise of 0.8 mb/d (or 0.9 per cent) on the previous year.
Whichever projection about how much oil is left turns out
to be true, it's unlikely that we will ever be able to increase
production much beyond 90 million barrels a day; a lot less
than we are predicted to need. Oil, regardless of how much
remains, is going to become a lot more expensive.

During 2012, the controversial process known as fracking, which liberates natural gas trapped within shale deposits,
picked up media coverage and is seen by many as a greener
alternative to petroleum and coal. The key point is that it
might well be 'greener', but it's not 'green'. Yes, natural gas
is the least carbon-intensive of the fossil fuels, but it's a very
long way from being carbon-free. Even switching exclusively

to natural gas, abandoning oil and coal altogether would do nothing to stop global warming and the effects of climate change, which would continue unabated, with devastating consequences for us all.

Clearly we need an alternative to oil to power our homes and industries and to fuel our trains, planes and automobiles, but our reliance extends much further than many of us think. The plastics we take for granted are all currently derived from petroleum, and it's at the heart of modern agriculture. That means that our alternatives to oil will have to do more than provide power and transportation; oil is used to create artificial fertiliser, so we need to dramatically increase the yield from land and produce crops that don't need it.

Finally, all these challenges must be dealt with against the backdrop of climate change. Extreme weather, rising global temperatures and rising sea levels will further impact food production and water supplies across the world. The areas likely to be hardest hit are those most important for food production: the mega-deltas of the Nile, Amazon, Ganges, Yangtze and other major rivers. The oceans, already over-exploited, will become less diverse as whole ecosystems vanish completely.

Can we weather the perfect storm? The answer is unequivocally yes, but in doing so we will create a world very different from the one we are living in today. Different – we can't paddle out of this creek in the same canoe we came in on – but not necessarily worse. Things should be much better for most people, and the lives we must lead will certainly be less wasteful. The one thing we should not fear is change. In helping to achieving this, the most plentiful resource we have is also the one that is most under-exploited. Our sun can give us all the energy we will ever need. As we noted above, if we harness all the energy in just one hour's worth of sunlight that reaches the Earth, we will be able to meet the planet's food and energy needs for an entire year. Thomas Edison, inventor of the electric light bulb, pioneer of alternating current and the power station, certainly recognised the opportunity: 'I'd put my money on the sun and solar energy', he said. 'What a source of power! I hope we don't have to wait till oil and coal run out before we tackle that.'

It's not beyond our ability to generate all the energy, food and water we need in a way that is affordable, sustainable and widely available. We know already how to increase crop yields five-fold and how to intervene so that waste can be practically eliminated. Nuclear and solar power can provide us with electricity, while biodiesel, harvested from vast third-generation photo-bioreactors connected to factories or on algal-diesel farms located in the world's barren deserts, can be used to fuel comprehensive public transport systems and private vehicles. But in order to understand fully how we are going to live in – and enjoy – the future, we need to understand how we got here in the first place.

3   Star Power

'We are just an advanced breed of monkeys on a minor planet of a very average star. But we can understand the Universe. That makes us something very special.'
Stephen Hawking, research director at the Centre for Theoretical Cosmology, University of Cambridge

The universe, and everything in it, was formed in an instant 13.7 billion years ago. Despite its incomprehensible size and age, the universe is neither infinite nor never-ending; it's a closed system to which nothing can be added, containing a finite amount of stuff. The Big Bang was not just the moment of creation, but the whole of creation as well. For that reason the entire history of the cosmos has been one of recycling and renewal. All the energy that we use on Earth, from the fuel we burn to the food we eat, is simply the latest link in a series of transitions and exchanges that began with the Big Bang.

That's not to say that it's been an undynamic or uneventful 13.7 billion years – and never more so than at the very beginning. During the first moment following its birth, known as the Planck epoch, the four fundamental forces that govern the universe – the weak and strong nuclear forces, gravity and electromagnetism – changed in nature, becoming disunified and independent, with gravity weakening significantly in the process. The energy from this separation drove the next phase transition – the equally fleeting grand unification epoch – which consisted of the aptly named cosmic inflation. Before it was a second old, the universe had grown from 10^{-25} metres to 10^{+25} metres in diameter, or from a size very much smaller than the nucleus of an atom to over one billion light years across (by way of comparison, the diameter of our own Milky Way galaxy is 100,000 light years). During the cosmic inflation, the universe was losing heat faster than a hot towel in a Chinese restaurant. Briefly this cooling allowed particles to form. So briefly, that by the time the universe was just one second old it was no longer hot enough or dense enough to create protons and neutrons, fixing their ratio for eternity. Four minutes later, the production of atomic nuclei was similarly frozen, leaving us with nothing but a whole load of radiation until the first generation of stars began to form some 380,000 years later. This early universe was unrecognisable from the one we live in today; a hot, dense soup of super-heated particles whizzing around in complete darkness.

It might appear that we know a great deal about the universe's formation and history, but it remains a very mysterious place. The observable universe accounts for just 5 per cent of its total mass. The 95 per cent that's missing is proving very difficult to track down. Scientists' current best guess is that over four-fifths of this elusive material is made up of dark matter. It really is a best guess too. Dark matter is completely undetectable, consisting of weakly interacting massive particles (WIMPs) – which, as their name implies, don't interact with anything – and massive astrophysical compact halo objects (MACHOs) that float through the universe emitting neither light nor radiation and enjoying no association with any visible stellar object. Scientists are searching actively for dark matter, but so far it stubbornly refuses to reveal itself. In July 2012, the latest results from the underground Xenon 100 detector in L'Aquila, Italy, showed no signs of a WIMP. We might not be able to find it, but we can feel dark matter's effects. This is the stuff that allowed galaxies to form and that is stopping them all from flying apart. But even if we accept the existence of dark matter, that still leaves around 17 per cent to account for. Scientists believe the rest is made up by the even more baffling dark energy. Dark energy's effects were first noticed in 1998 when scientists discovered that something was overpowering gravity and causing the universe's rate of expansion to increase. Beyond that we don't really know much about it at all.

Returning to what we can see: when we look up at the night sky, what we find are stars – billions upon billions of them. Stars come in a spectacular range of colours, sizes and brightness, but they are all made up of just two elements: hydrogen and helium. Together hydrogen and helium account for almost all of the visible matter in the universe: some 97 per cent. All the things that seem so common to us here on Earth – nitrogen, oxygen, iron, water, gold rings, *X-Factor* contestants and Tesco – are actually very rare indeed. Stars burn brightly because they harness the power of nuclear fusion. This is a different kind of nuclear reaction to the one we use in power stations, submarines and cruise missiles. That method is fission, and it works by splitting atoms apart to release vast amounts of energy. In stellar fusion the opposite happens. Deep in the core of a star temperatures are very high, causing hydrogen atoms to fly around so quickly that they collide and fuse to form helium. These helium atoms have slightly less mass than the two hydrogen atoms that created them. The mass that is lost escapes in the form of radiation.

Energy and mass should be viewed as two sides of the same coin, rather than two different things. All objects are moving through space. Even when you're lying in bed, the Earth is still travelling at 67,000 miles an hour on its orbit around the sun, while the whole solar system is orbiting the galaxy's core at 514,000 miles an hour and even the Milky Way itself is hurtling through space. When you add it all up, it means that we're moving at around 3,240,000 miles an hour even when we're asleep. We don't feel like we're moving, because as Einstein pointed out, speed is relative. Similarly, when we're travelling in a train, car or aeroplane we don't feel like we're moving at all, unless there's an increase or decrease in speed. An object's mass is simply a measure of its resistance to the change of speed. It takes much more effort to get a boulder to start rolling down a gentle hill than it does a pea, but the boulder is also much more difficult to stop once it's moving.

The force employed to get objects to change their speed is energy. Physicists define energy as the ability of any closed physical system – like an atom or a star; a cell or a tree – to 'do work' on any other physical systems. This work takes the form of a force that acts over time and space, the equivalent of a pull or push against the four fundamental forces of nature. These forces act upon everything all the time. Even as you sit comfortably reading this book, the force of gravity is what's pulling you down into the seat and the force of electromagnetism is responsible for generating the light by which you are reading. The two remaining forces work at an atomic level and can't be felt, but are just as important. The strong nuclear force is what holds atoms together: without it, you and the book you're holding wouldn't exist – nothing would. The weak nuclear force – sometimes called weak interaction – is responsible for radioactive decay and is what initiates the process of hydrogen fusion in the heart of our sun. No weak nuclear force means no fusion, which means no stars, so no sun and, ultimately, no life on Earth.

Einstein's famous equation $E = mc^2$ (energy = mass × the speed of light squared) shows us how energy and mass are linked. The total energy contained in any object – a hydrogen atom, a lump of coal or a chocolate muffin – is identified with its mass. Like mass, energy cannot be created or destroyed, merely converted into one form or another. When matter is converted into a form of energy – for example when the chemical energy of petrol is converted into kinetic energy within an internal combustion engine; or when the binding energy of two hydrogen atoms is converted into radiation during fusion – the mass of the system does not change throughout the transformation process. In fact, if you could measure it accurately enough, you would find that a freshly baked loaf of bread weighs more than an identical loaf that has cooled down, because it has more energy.

※

Energy cannot often be measured absolutely. Only its

transition from one state into another can be calculated, and often only in relative terms. The conventional method of measurement is calorimetry, which is based on temperature or the intensity of radiation. The energy transferred is expressed in a variety of units such as ergs, calories, joules or – in the UK especially – kilowatt hours (kWh). The usefulness of each unit depends very much upon what is being measured, but they are all interchangeable. For example, 1kWh is equivalent to 3.6 million joules. The former is useful for measuring domestic energy use and calculating electricity bills; the latter for determining the energy content of food.

When it comes to discussing fuels, a much more helpful measure is energy density, which refers to the amount of useful energy stored per unit volume or per unit mass. For example, when comparing the effectiveness of hydrogen to petroleum as fuels, if we look just at the specific energy contained in each, then we will see that hydrogen contains far more energy pound for pound. However, in practical terms, petroleum is much more effective fuel because, even in liquid form, hydrogen has a much lower energy density – effectively providing fewer miles per gallon.

Energy can take several different forms. It can be chemical, nuclear or magnetic; mechanical, radiant or thermal. There are other kinds too, but all these forms of energy can be more practically divided into just two groups: kinetic energy – the energy an object possesses due to its motion; and potential energy – the energy contained within a physical system. A car moving down a road has kinetic energy (produced by converting the chemical energy of the fuel into thermal energy that drives the pistons to produce mechanical energy), while the diesel in its gas-tank has potential energy (chemical energy that can be burned in an engine to produce thermal energy, etc.). Energy can be converted from any one of these forms into another: sometimes with nearly 100 per cent efficiency, but often with much less, as some of the energy leaks away as heat.

The c^2 bit of Einstein's equation also shows us that tiny amounts of matter can produce huge amounts of energy, because the speed of light squared is a very, very big number. Written down, it looks like this: $8.98755179 \times 10^{16}$ m²/s². This is why fusion is such an efficient means of producing energy. In the heart of our own sun, every single second 600 million tonnes of hydrogen collides to create new helium, releasing 4 million tonnes of energy in the process. Yet this has already been happening for almost 4.6 billion years and will continue for several billion more. If the sun produced the same amount of energy by burning coal instead, then its lifespan would be very much shorter. Burning is the conversion of molecular bonds, rather than sub-atomic binding energy, into radiation. Burning coal yields 25 kJ/g (kilojoules per gram) by breaking carbon-to-carbon and oxygen-to-oxygen molecular bonds and forming carbon-to-oxygen bonds. This is not very efficient when compared to converting mass directly into energy, which yields nearly 4 billion times more: 8.988×10^{10} kJ/g. A 3kg bucket of coal is good for about 20 kilowatt hours – enough to heat a small house for about an hour – and releases energy at a rate of 7 kW, that is 7 kilojoules per second. The sun's mass is 2×10^{33} kg and it releases its energy at the rate of 4×10^{23} kW, so if it was burning coal instead of converting mass into energy, it would last for only 4,000 years rather than several billion.

This is also why fusion is the most common form of energy production in the visible universe: it's able to power every star for billions of years. But long though their lifetimes are, stars don't last for ever; just like us, they are born, live and die. How long they live for and how they die depends on what kind of star they are. Our own sun is an unremarkable G-dwarf star. As it grows older it will become smaller, brighter and hotter. In three billion years' time, it will be 40 per cent hotter than it is today, evaporating all the water in our oceans and bringing an end to all life on the planet. In five billion years' time, its core will collapse as it runs low on fuel and it will expand to become a massive red giant, engulfing all the inner planets – Mercury, Venus, Earth and Mars

– in the process. Eventually it will throw off its outer mantle in a mass of gas and dust, leaving behind a white dwarf to twinkle over trillions of years. Not all stars bow out so bathetically. Big stars live fast, burn brighter and die younger in supernovae.

During their lives, stars provide the universe with light and heat, but their deaths are just as productive. No one knows for sure how long an atom can survive, but they are remarkably durable, with conservative estimates putting their lifespan at 100 billion trillion years. Across the universe stars are dying all the time; but the night sky doesn't get dimmer because everything is recycled and new stars are also being formed all the time. The atom is at the heart of this cosmic recycling scheme. Stellar nucleosynthesis is the process that creates new atomic nuclei from existing protons and neutrons within stars, and it's how we managed to get all the other primordial elements after hydrogen and helium from carbon to iron. As stars get older and hotter, fusion starts to produce heavier elements from the hydrogen and helium nuclei. All the elements that are heavier in atomic number than iron, up to plutonium, are produced by explosive nucleosynthesis, which occurs during a supernova. Recycling begins as soon as the dying star explodes, leaving clouds of superheated atoms and particles. Gravity causes these clouds to condense. As the matter becomes compact, so it heats up and eventually becomes so hot that it ignites and a new star is born.

This process has been going on almost as long as the universe itself. Our sun is a third-generation star, which means that every atom in our solar system has been through this process twice before. The solar system began its life as one of these clouds of cosmic dust called the Solar Nebula. About 4.6 billion years ago this nebula coalesced under the force of gravity and began spinning into a flattened disc. 99 per cent of all the material was compressed into a proto-sun – not yet ready to ignite – while the rest formed rings. Close to the sun only rock and metals could survive the heat, which is why the planets from Mars inwards are made of heavier material, predominantly iron. Further out, big lumps of rock and ice were able to attract gas, which is why the outer planets – Jupiter, Saturn, Uranus and Neptune – are made of much lighter stuff.

Eventually the sun became hot enough and dense enough to ignite, but the ancient solar system was still a chaotic place and astral collisions were common. When the Earth was just 100 million years old it collided with another proto-planet called Thea, which was about the same size as Mars. The collision knocked the Earth 23.5° off its axis, giving us the seasons, which provided some of the stimulus for evolution to take place. It also increased the velocity of rotation, which reduced the daily variation in temperature and changed the surface composition of the planet, enriching it in silica and other minerals. Most importantly of all, though, it formed the moon.

The Earth is unusual because it's the only inner planet with a significant satellite. Neither Venus nor Mercury has a moon, while Mars' twin moons Phobos and Deimos are nothing more than tiny rogue asteroids that became ensnared by the planet's gravity. Our large moon – approximately 27 per cent of Earth's size – confers a number of essential benefits. Without it the ocean tides would be only half as big, so the planet would lack the tidal pools that many scientists believe are vital for the formation of complex life. It's likely that the moon also initiated the process of plate tectonics, which are thought to be similarly essential, because they allowed the replenishment of nutrients for the primitive life-forms to feed on. Certainly, there is no evidence of plate tectonics on Venus, Mars or Mercury – and probably no life either.[1]

The Earth orbits the sun in the centre of what astronomers and astrobiologists refer to as the habitable zone; a

---

[1] Although in the case of Mars, we wait for the results of the ongoing Mars Rover Discovery Mission to see whether there is any evidence for water on the Red Planet, or even that it once contained life.

place where a planet with sufficient atmospheric pressure can maintain liquid water on its surface. The water on Earth probably arrived in the early part of its history as a result of collisions with stray comets and other icy celestial bodies, but however it got here it was this water that allowed life to begin and flourish. The Earth was also big enough to hold on to its atmosphere. On smaller planets, gravity is insufficient to hold on to much of the surrounding gases. Planets without an atmosphere are prone to enormous changes in temperature. On atmosphere-less Mercury, the temperature fluctuates between 400°C during the day and −200°C at night. This diurnal variation makes it impossible to sustain substantial oceans. Small planets also tend to have rougher surfaces with larger mountains and deeper canyons: were the Earth to be scaled down in size, it would be smoother than a billiard ball.

It seems that a set of coincidences made all the conditions on Earth 'just right' for life. This observation is known as the 'Goldilocks Enigma' and is the cause of much debate among scientists. Some people believe that the universe itself has been fine-tuned for life to exist, which certainly appeals to those looking for evidence of an intelligent designer. Others point out that if conditions weren't 'just right' for life, then there would be no one here to observe it – the so-called anthropic principle. Whether by luck or design, as far as we know this is the only place where life does exist, but how it came to exist is yet another mystery. Evidence suggests that there has been life on earth for 3.7 billion years, but there is no standard model regarding its genesis. Almost all of the competing models concur with the Oparin–Haldane hypothesis, which posits that a primordial soup of organic molecules could be created in the oxygen-free atmosphere of the juvenile Earth through the action of sunlight. However, beyond this basis there is a range of plausible but unproven ideas and theories based on disparate discoveries and a healthy portion of conjecture.

What constitutes life is another grey area. Living things

are distinguished by the fact that they have signalling and self-sustaining processes, unlike inanimate objects or those in which these functions have ceased (the dead). The earliest and simplest life-forms are the prokaryotes; a group of tiny organisms that lack a nucleus. They include the bacteria that we are familiar with, but also a whole host of organisms called archea, which are superficially very similar but have a completely different membrane and actually sit in a different family. Probable fossils have been found dating back almost 3.5 billion years. Viruses are even smaller than prokaryotes, and possibly even older (their origin is unclear because they don't fossilise); certainly they are found wherever life exists. At the fringes, though, opinion is divided about whether something is alive or not. There is no consensus as to whether viruses are a form of life or just organic structures that can interact with living things. In common with living things viruses have genes, and reproduce via natural selection and by making multiple copies of themselves, but they do not have a cellular structure and lack a metabolism – the set of chemical reactions that takes place in living organisms to sustain life – and must rely on a host to do that for them.

One theory that doesn't concur with Oparin–Haldane is that of 'Panspermia'. Popularised by the English mathematician and astronomer Fred Hoyle, Panspermia rejects earth-based abiogenesis (the creation of life from non-living matter), proposing instead that life exists throughout the universe and is distributed by meteors, comets and asteroids. What we do know for sure is that the creation of life must be a pretty rare event, because as far as we can tell it has happened only once, and no one has yet been able to synthesise even a proto-cell in the laboratory. We may never know the truth, but we are a living and conscious testament that cellular carbon-based life 'as we know it, Jim' did emerge on Earth some 3.7 billion years ago. And it has thrived ever since.

For all its variety and verdant splendour, life is in essence nothing more than a highly complex process of energy transfer. The biological conversion of energy is much less efficient than the fusion taking place in the middle of the sun, but it yields arguably even more spectacular results. The cells that make up all living things are life's basic functional units, but also the vectors for all energy transfer. Cells are tiny bags of plasma surrounded by a membrane, and each human body contains about 10 trillion of them. The cell membrane is one of evolution's *tours de force*. Membranes are semi-permeable, which means they not only protect the interior of the cell from the outside world, but also control the flow of substances into and out of it. Membranes are made of fat-like molecules called lipids, and embedded within them are a variety of proteins. Inside the cell is where all the action happens. A jelly-like substance called cytoplasm holds all the important bits in place. Genetic material in the nucleus – DNA and RNA – gives the cell its instructions, determining what kind of cell it is – skin, bone, blood, etc. – and what kind of job it does – store fat, fight infection, and so on. The actual work is done by the organelles, a set of miniature organs: ribosomes are factories that process amino acids, centrosomes organise the cell and maintain its structure – the cytoskeleton – while vacuoles store food and get rid of the waste.

The most important organelles from our perspective are mitochondria, and in plants, chloroplasts. Chloroplasts make carbohydrates, the basic fuel for all life-forms, while mitochondria burn carbohydrates to make ATP (adenosine triphosphate): the energy vector that powers the processes of life. Mitochondria generate much of the cell's supply of chemical energy by converting sugars, fatty acids and nucleotides into ATP. ATP is sometimes called the 'molecular unit of currency' because it's the vehicle by which chemical energy is transported for use by the metabolism. The metabolic processes use ATP as an energy source by converting it back into its component parts. ATP is therefore continuously recycled in organisms. For example, the human body, which on average contains only 250 grams of ATP, turns over its own weight in the substance each day. Chloroplasts are found only in cells of plants, algae and cynobacteria – they are also what gives these life-forms their distinctive green colour. Like mitochondria, chloroplasts are responsible for producing ATP, but they do it by capturing and converting light energy through photosynthesis. In photosynthesis, carbon dioxide is converted into organic compounds – such as sugars – using the energy from sunlight, and releasing oxygen as a waste product. Photosynthesis is vital for all aerobic (oxygen-breathing) life on the planet. It regulates the levels of oxygen in the atmosphere and is also the primary source of energy for almost all life on the planet, whether directly – as in the case of plants – or indirectly as the ultimate source of the energy in food – as in the case of things that eat plants and things that eat things that eat plants. There are a small number of exceptions – biosystems that have formed around deep-sea hydrothermal vents, for instance – but as a general rule of thumb, it's helpful to think of all food as simply captured sunlight.

The earliest photosynthetic microbes lived in the ancient oceans surrounding the singular continent of Pangaea. There was no life on the land at the time, but these organisms transformed the atmosphere by excreting oxygen as a waste product. In the early stages, this oxygen remained water-bound, where it converted iron II into iron III. Iron exists in two oxidation states called II (ferrous oxide) and III (ferric oxide); the former is water-soluble, the latter we know as rust. Only once all the iron II had been oxidised could the oxygen begin to build up in the atmosphere, eventually reaching, after a couple of fluctuations, the homeostasis of 21 per cent we know today. Photosynthetic microbes are also responsible for creating all the iron ore that mankind has used to transform the Earth. It's worth considering that the coal and iron ore that drove the industrial revolution are merely relics of a much earlier energy revolution involving photosynthesis. Certainly, we won't find coal or iron ore on dead planets.

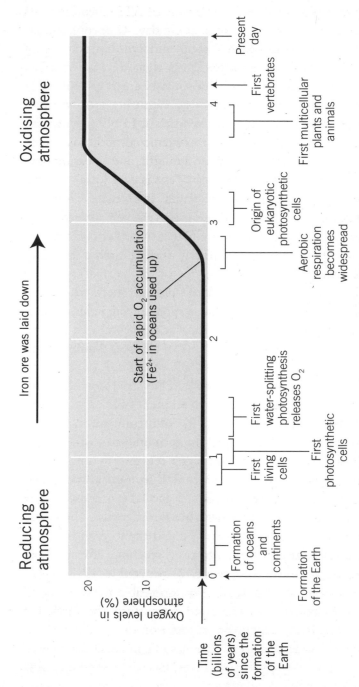

Fig. 2: Oxidation of the Earth's atmosphere with the key stages of evolution noted.
(Source: redrawn from *Teaching About Evolution and the Nature of Science*, National Academy Press)

Photosynthesis provides us with more than just food. Its rate of energy capture is immense, approximately 90 terawatts (TW) per year, which is about six times larger than the current power consumption of the whole of human civilisation. Photosynthesis is also the source of the carbon in all the organic compounds within the body of every organism. In all, photosynthetic organisms convert 115 billion tonnes of carbon into biomass every year. Photosynthetic organisms are 'photoautotrophs', which means that they are repositories of energy. What isn't consumed as food can often be preserved long after the plants or algae themselves have expired. When we burn a log, it's this repository of stored energy that we are tapping into, and the same is true when we burn a lump of coal. The only difference is that the energy in a lump of coal is much older.

Coal is the most abundant of the fossil fuels, which also include petroleum and gas. These fuels are the fossil remains of dead plants and animals whose stores of cellular energy, built up while they were alive, remain intact. These organisms expired long ago, typically tens of millions of years and in some cases well in excess of 650 million years ago. Because they were once living things, fossil fuels contain high percentages of carbon. Fossil fuels formed from the remains of organisms fossilised by exposure over millions of years to heat and pressure within the Earth's crust. In the case of coal these organisms were the trees that, 300 million years ago during the Carboniferous period, formed dense forests in low-lying wetland areas, which covered almost the entire globe. The wide, shallow seas of this period provided ideal conditions for coal to form. Natural processes like flooding caused these forests to get buried under the soil. A combination of mud and acidic water prevented decomposition from taking place and instead the trees were covered with increasing layers of soil, sinking deeper and deeper into the Earth. Before long, all the carbon became trapped in immense peat bogs, which were eventually covered and buried by further sediments. Over millions of years, the pressure and temperature increased and the vegetation slowly became converted to coal – a process called carbonisation.

Unlike coal, petroleum and natural gas are formed by remains of algae and zooplankton that lived in prehistoric oceans. When these creatures died, their bodies settled in large quantities at the bottom of seas and lakes, where oxygen was scarce. This absence of oxygen prevented normal decomposition from taking place, preserving the organic matter, which in the same manner as coal became buried under layers of sediment. Over millions of years heat and pressure built up, causing a chemical transformation into a waxy material called kerogen, which is found today in the oil shales of Brazil and Canada. However, much of this kerogen was subjected to further heat and pressure, which turned it into the gaseous and liquid hydrocarbons familiar to us as natural gas and petroleum. This process is known as catagenesis.

Burning fossil fuels releases not just this ancient energy, but also carbon dioxide: around 22 billion tonnes every year. Plants do absorb carbon dioxide, but together with other natural processes they can deal with only around half of that amount, which means that each year there is a net increase of 11 billion tonnes of atmospheric carbon dioxide. Carbon dioxide is one of the greenhouse gases that contribute to global warming, which is causing the surface temperature of the Earth to rise and producing major, adverse effects for us all.

Fossil fuels are not renewable, because they take millions of years to form and reserves are being depleted much faster than new ones are being made. Indeed, it's worth considering just how vast these timescales really are. Begin by counting to 100 in one minute (if you can't be bothered, just take our word that it's an easy feat to achieve). Now let's imagine each number is equivalent to a year. By the end of just one minute, we have already reached a point before the birth of most of the people on the planet. Continue counting at the same rate, and 20 seconds later we will have arrived at a point in history before even today's most venerable senior citizens were born. By the end of the fourth minute we are at the start of the industrial age. The birth of Christ takes place after 20 minutes and at 50 minutes we come to the dawn of civilisation, which began in small cities on the banks of the Tigris and Euphrates. It will take us an hour and 20 minutes to reach the dawn of farming. If we continue without stopping for a week and a half, we will be at a time when the earliest hominids started to use tools. To get to the time when the most recent deposits of coal were still trees will take nine months, and if we carry on until the first coal was laid down, we need to keep counting for another twelve years. In all, it has taken over 600 million years to produce all the coal, petroleum and natural gas in the world, and we are currently burning through around 20 million years of this fossil record every single year.

The development of cells that could photosynthesise was the first effective mechanism for capturing sunlight, while the process of fossilisation has ensured that the fruits of this process could be stored almost indefinitely until harvested by mankind to provide heat and power. Like the food we eat, the oil, coal and gas we burn with abandon is nothing more than captured sunlight: a single process of energy transfer, albeit one punctuated by millions of years, which began with the Big Bang, continued through the nuclear fusion reactions of three generations of stars and the photosynthesis of ancient plant life, and ended up as fuel for the daily commute and coal for the fire.

In this context it's easy to see how precarious is our reliance on fossil fuels. Regardless of how much is actually left, and what damage its ongoing exploitation will do to the environment, we'd do well to recognise it for the rare and precious resource it is. There may yet remain mountains of coal and seas of oil hidden beneath the earth, but this is certainly the only place in the solar system where we will find any sources of trapped sunlight. Using them with less wanton abandon can surely be no bad thing – consider the thousands of freight-carriers that are, even now, making their way to European and American ports from China and India, laden with precious cargoes of Christmas cracker gifts, mobile phone covers, mouse mats, key fobs and other indispensable trinkets and geegaws.

Oil, coal and gas: we'll miss you when you've gone. But go they most certainly will.

4   Taking Control

> 'Man is the only creature who refuses to be what he is.'
> *Albert Camus, philosopher*

Physiologically speaking there is nothing special about human
beings. Our species *Homo sapiens* is the only surviving member of the Homo genus today, but as recently as 10,000 years
ago we probably shared the planet with at least three others:
Homo floriensis, Denisova homonin and the so-called Red
Deer Cave people, whose fossilised remains were discovered
in China during 2012. *H. sapiens* probably emerged around
200,000 years ago, but if we measure evolutionary success in
terms of longevity then we've got some way to go before we
match the 1.6 million years endured by *Homo erectus*. Even
the big brains we take so much pride in are by no means
unique; adult Neanderthals (*Homo neanderthalis*) had brains
at least the size of ours, possibly larger – and so do bottlenose
dolphins and orcas.

    Yet we are special. Mankind may be the only animal that
hunts for sport,[1] but we are also the only animal that plays
Xbox LIVE, goes on holiday and makes chocolate brown-
ies. There are innumerable differences between *Homo sapiens* and the rest of the animal world. Our lives have little
in common even with our closest relatives, chimpanzees,
bonobos and gorillas – that is immediately apparent. Yet perhaps the biggest difference of all is not how we live, but how
we end our lives; mankind is the only species afforded the
luxury of dying from old age. The average person can reasonably expect to live way beyond the two score years that
was the lifespan of our most venerable prehistoric ancestors
and though it's an unfortunate fact of life that we will probably have to endure a period of serious illness at some point,

---

[1] Although cat-owners may beg to differ.

many of us will make a full recovery. The same cannot be said for tigers, squirrels, beetles, blue whales, orang-utans, Neanderthals or anything else you care to mention. The life of almost every non-human animal living in its natural environment ends prematurely and in almost all cases, horribly.

Prior to the publication of Darwin's *On the Origin of Species* in 1859 the view of nature, widely held and rarely challenged, was of a carefully designed, unchanging system wherein every living creature had its own special place. This notion of God as the designer of the natural world was best articulated by William Paley in his 1802 book *Natural Philosophy*, which posits the idea of a 'Divine Watchmaker'.

The possibility of extinction was not widely accepted before the 19th century. People believed instead that a great chain of being extended from God, at the apex of course, down through the angels and demons, to men and women, with the animals, plants and minerals at the bottom. It had been ever thus and so it would remain for ever. Darwin's theory of evolution shattered this prevailing uniformitarianism, but what the Victorians found most shocking of all was the revelation that the state of nature really wasn't a very nice place at all. Those 'bright and beautiful' animals were all living in constant fear, if not abject terror, because murder, starvation, pain, disease and suffering were daily hazards of normal life – and untimely demise an absolute certainty.

Today, we tend to think of ourselves if not exactly outside of evolution, then at least invulnerable to its more unpleasant aspects. But we are as much a product of its processes as lice and lions; dust mites and dinosaurs. Extinction is a real and present threat. It is estimated that 99.9 per cent of all the species that have ever existed are now extinct. Evolution leads to new species arising through the process of speciation, where modified varieties of existing organisms thrive because they are able to exploit a particular ecological niche. A species becomes extinct when it's no longer able to survive in changing circumstances against better-adapted competition. Typically, a species becomes extinct within 10 million years of its first appearance, although there are exceptions where creatures survive virtually unchanged for hundreds of millions of years. Coelacanths, horseshoe crabs and nautiluses are examples of these so-called 'living fossils'.

Most extinctions that occurred prior to the emergence of *Homo sapiens* were due either to one species being outcompeted by another or to changes in the environment. The disappearance of one species creates opportunities for another, but when many disappear at once – in a mass extinction – the rate of speciation increases dramatically, with many new species emerging at the same time. These mass extinctions are extremely rare events. In the past 540 million years there have been just five occasions when large-scale plant and animal extinctions took place over a geologically short period of time. The most recent was the Cretaceous–Paleogene extinction event (or K-Pg for short), which occurred some 65.5 million years ago and accounted for an estimated 70 per cent of all life on Earth. Scientists believe that K-Pg was caused by one or more natural catastrophes: at least one asteroid impact or increased volcanic activity (perhaps caused by the former) that released masses of dust and ash into the atmosphere, radically altering the global climate and reducing the ability of plants to photosynthesise. Most animal life was unable to come to terms with these sudden environmental changes, most notably the non-avian dinosaurs. In the oceans, the ammonites, a genus of marine invertebrate that had dominated the seas for almost half a billion years, also disappeared.

The animals that did survive were those more able to adapt. Almost all the mammalian species that flourished at the end of the Cretaceous period were small – about the same size as rats – and many lived semi-aquatic or burrowing lifestyles that would have increased their chances of survival. Evolutionary success depends as much on fortune as fortitude. Species are very much a product of the environment, but the laws of physics determine the size to which plants and animals can grow. Arthropods (insects, spiders, crustaceans) are as large as they can get based on the limits of their respiratory and circulatory systems, which work by diffusion. Because they don't have lungs and circulatory systems like mammals, they simply could not supply their tissue with enough oxygen if they got any bigger. During the Carboniferous period there was a good deal more oxygen in the atmosphere, hence 30cm-long spiders and dragonflies with 1-metre wingspans were not unusual. Similarly, at the opposite end of the scale, mites have reached the lower limit of possible size for arthropod design – they can't get any smaller and metabolise properly. Plants have also reached the limits of how big they can get, because the vascular tissue that carries water and nutrients from root to leaf tip is subject to both gravity and pressure differences within the cells of the plant. Thus, the giant redwoods in California are about as big as trees can get. The reason dinosaurs were bigger than mammals is that these giants were cold-blooded and required less energy input to maintain their metabolisms. A mammal equivalent in size to one of the giant sauropods would be possible in terms of physics, but due to it being warm-blooded, such an organism would require more than ten times as much food and would need to spend its entire life eating. Whales benefit from the buoyancy of their watery habitat, which offsets the gravitational stress on bones and joints they would experience on land. They grow so large because their food sources – like plankton, krill or squid – are so plentiful and easy to obtain. *Homo sapiens* is a medium-sized hominid; if even our strongest representative were to encounter an average Neanderthal, *Homo heidelbergensis* or *Homo erectus* on a dark night, the ensuing fight would be over almost immediately and terminally, but we could all muscle our way to the front of any queue consisting of *Homo habilis* or *Homo floriensis*.

※

Just as the conditions were 'just right' for life on Earth to emerge 3.7 billion years ago, so the K-Pg was necessary to create the right conditions for mammalian life to flourish and for intelligent life to evolve. It wasn't only mammals that flourished after K-Pg. Grasses – including cereals and bamboo – have played a vital part in human evolution, and they too became widespread only after the Cretaceous period. In the millennia following K-Pg, the world was dominated by birds – descendants of the avian dinosaurs – and ancient groups such as the crocodiles. Grasses changed the landscape, with the global forests of the Carboniferous giving way to steppe and savannah by the time of the Miocene, 25 million years ago. This is the change that allowed mammals to finally dominate. Species of grazers appeared – occupying the same niche and similar to modern cows and horses – as did the animals that preyed upon them, the carnivora: cats, dogs and the now-extinct creodonts (like patriofelis, which looked a bit like a cross between a cat and a dog). The evolution of primates goes back at least 65 million years and possibly even to the mid-Cretaceous period some 85 million years ago, but for our own species the emergence of the grasslands was pivotal.

Biologists have spent over 100 years discussing the significance of the savannah for human evolution, and at various times upright walking, tool-use and increasing brain size have been attributed to these grasslands. Our earliest human predecessors diverted from a common ancestor with forest-dwelling chimpanzees around 6 million years ago, at a time when the African savannah was prevailing against woodland and dense forest. In order to survive in this new environment, where protection through tree-cover was 80 per cent less than in the forest, it obviously paid to be able to see where you were going and what predators might be hiding out there. Bipedalism also conferred other benefits, freeing the hands for grasping objects or for carrying food and young. Fossil evidence now shows us that bipedalism far pre-dated large brains. The first primates to walk upright were the australopithecines, a genus that evolved in East Africa around 4 million years ago containing at least nine separate species.

It is widely held that *Australopithecus* eventually evolved into the first member of our own Homo genus *Homo habilis* (literally 'handy man), but their brains were no more sophisticated than modern-day chimps or bonobos.

These advantages of bipedalism might have been slight, but over thousands of years they conspired to reward the smarter, more inquisitive members of the species with survival. This process of successful adaption is evident in *Homo habilis*, who lived between 2.33 and 1.4 million years ago in Eastern and Central Africa. *H. habilis* was about half the size of modern humans, but with a less protruding 'ape-like' face than the australopithecines, and possessed the opposable thumbs that separate hominids from other primates. Opposable thumbs helped *H. habilis* to develop more accurate fine motor skills (it's likely that teenaged *H. habilis* would have been dexterous enough to incessantly text their friends using a mobile phone; not so adolescent chimps and bonobos – how do they cope?). *H. habilis* was also more intelligent than any species that had come before, with more sophisticated social organisations than either australopithecines or chimpanzees, but we should not be fooled into thinking he was as yet master of all he surveyed. Despite a strong claim to his being our earliest 'human' ancestor, the fossil record also reveals that this early hominid was the staple diet of several species of large sabre-toothed cat also living at the time.

Homo erectus, possibly a descendant of *Homo habilis*, was the first species of human to start controlling the environment. There are indications that *H. erectus* might have been capable of controlled use of fire as long as 1.5 million years ago, but firm evidence dates back only 400,000 years (and for the widespread use of fire, we can find evidence from only 125,000 years ago). This management of fire is just as significant a development in our evolution as tool-making or bipedalism. Fire allowed activity to continue into the night, providing warmth, light and protection from insects and animals. Fires do occur naturally, but usually with dire consequences for any animals curious enough to investigate them up close. Chimps and other apes are scared of fire, and it's reasonable to assume that our early ancestors would have found it difficult to overcome the same natural inclination for flight. It's likely that a long period of opportunistic – and highly dangerous – employment of fire preceded any attempts to manage it, let alone create it. Making a fire from scratch is not easy to do – a series of failed merit badges from our scouting days are testament to this – nor is it intuitive; it requires an understanding that friction will, in the right circumstances, produce a flame. We should never forget that this skill took us hundreds of thousands of years to master.

As well as providing light and heat, fire also allowed us to cook food. This was arguably the most important long-term benefit. Cooking food, especially proteins, improves nutrition, making it more digestible and allowing us to absorb more calories. Cooking also increased the amount of potential food available, because it helps the body to break down the starchy carbohydrates in leaf stems, roots and tubers that were not previously part of the hominid diet. That early diet, consisting of nuts, berries, fruit and raw meat and vegetables, would not have provided enough nutrition to support anything more than extended family groups.

In *Homo sapiens* we see other results of this culinary success. Cooking food means that it's partially digested before it reaches our mouths, so in comparison with apes we require smaller mouths and stomachs, weaker jaws and a 50 per cent shorter large intestine. Author Richard Wrangham tried living on the same diet as chimpanzees without much satisfaction. Appetising though many tropical fruits appear, they lack the nutritional value of root vegetables, while the ape's favourite flavour – mustard oil – caused his tongue to freeze in the manner of a novocaine injection. Cooked food takes less time to chew. We spend no more than an hour each day eating, while chimps are grinding their teeth for six.

Whether these improvements in diet had a significant impact on brain size is a moot point. The mainstream view is that human brain size increased long before the advent of cooking, due to the increased consumption of meat, but there is evidence that cooking was critical in the development of language, for sociological as well as physiological reasons. *H. erectus* was the first hominid to live in hunter-gatherer societies, and socially they had much more in common with us than with *Australopithecus*. Effective cooperation within these groups would have required some form of communication and it seems likely that language developed over a long period of time, rather than arriving fully-formed. Indeed, it's hard to imagine the sophisticated tool-making, fire-lighting, cooking and collaborative hunting practised by *H. erectus* and its descendants *Homo ergaster* and *Homo heidelbergensis* taking place without any communication at all. More likely they communicated using a proto-language, which may have lacked the syntax and structure of true speech but would have been much more sophisticated than the non-verbal communication used by chimps. The morphology of the middle and outer ear supports this view, suggesting that the auditory sensitivity of *H. erectus* was much like that found in humans and very different from modern apes.

The development of language was the final evolutionary step on the way to our modern way of life. *H. sapiens* was the first, and as far as we know the only species to develop fully formed speech. While there is evidence that Neanderthals developed their own sophisticated form of language – albeit one that would have sounded very strange to our ears – it was slower-paced and they appear to have been incapable of articulating complex ideas. This failure to achieve a comparable level of verbal communication may well be what ultimately did for the Neanderthals as a species.

Homo sapiens and Neanderthals coexisted for around 100,000 years. Neanderthals where quite different physiologically from *H. sapiens*. They were better adapted biologically to cold weather; shorter-limbed, barrel-chested, they were also stronger and more robust and they looked very different too with their wide, flat noses and protruding brow ridges.² Neanderthals lived in colder climes, roughly along the 50th parallel north, but south of the line of glaciation during the last ice age. Their remains have been found in most of western Europe; in the east as far as Siberia and around the north Mediterranean. To date there have been no findings in Africa, where *H. sapiens* first appeared. The remains of both species have been found at the same site in the Middle East, where it seems the Neanderthals arrived after *H. sapiens* had vacated.

But despite their bigger brains, they were not necessarily more intelligent. They were resourceful, certainly more so than they've been given credit for in the past, and their tools are now thought to have been as good qualitatively as those made by *H. sapiens* living at the same time. But for some reason they failed to develop equally sophisticated survival strategies. Neanderthals never 'invented' arrows or throwing spears, which meant that their hunting technique – ambushing big game and going in for the kill at close range – was far more likely to result in death or serious injury than that of *H. sapiens* who, with their bows, arrows and other missiles, could bring down prey from a safe distance. Their larger size also meant that the average Neanderthal male required up to

---

² Despite many competing theories, the real reason for this contrasting physiognomy is most likely a minor accident of genetics. The Neanderthal population, even at its peak, was very small – not more than 70,000 individuals. Face shape confers no real survival benefits in a colder climate. Anthropologist Chris Stringer investigated the competing hypotheses for the Neanderthals' strong brow ridge and concluded that none is particularly convincing. The ridges are hollow, so they are not there to transfer a physical force (in the form of, say, a 'Glasgow handshake'), nor do they help heavy chewing. In one experiment a scientist strapped on a replica brow ridge to investigate its possible advantages. It shaded his eyes from the sun, kept his hair out of the way when running and frightened people senseless on dark nights. Perhaps this last point is most significant: they could have been a signalling tool, accentuating aggressive stares, especially in males, with larger sizes sexually selected through the generations like antlers in deer.

350 more calories per day to survive than *H. sapiens*. This is a significant difference – around 10 per cent – which would become a major disadvantage at times when food was scarce or competition for resources was fierce.

☀

There are few disadvantages to being a hunter-gatherer, which explains why it was the hominid subsistence strategy of choice for almost 1.8 million years. Anthropologists used to believe that being a hunter-gatherer was a precarious existence, but recent studies have shown that they spend a lot less time collecting food from various sources than equivalent agricultural communities spend growing it. With a thorough knowledge of the available resources in an area, it's possible to predict with relative certainty when and where to find food. The effects of a bad season would be felt by both farming and hunter-gathering communities equally. In fact the main disadvantage of hunter-gathering is, like all lifestyles, that there's an upper limit on the number of people it can support. Less established, smaller or weaker groups are pushed into more extreme environments. The Neanderthals, relatively inarticulate straight-line thinkers living in smaller communities, were vulnerable to this process and ill-equipped to develop alternative survival strategies, such as farming. Whether some of their species were assimilated into the *H. sapiens* population or whether they disappeared bathetically due to climate change, or perhaps in an early instance of ethnic cleansing, we will probably never know for sure. But by 24,000 years ago there were none left.

For *H. sapiens* the shift from hunter-gathering to farming was a gradual process. In the middle to upper Palaeolithic period (80–70,000 years ago) we find specialisation beginning to occur, with communities focusing on a smaller selection of game and gatherings. In coastal areas, bespoke tools such as fish hooks, bone harpoons and nets are developed, while elsewhere food production is becoming more organised through a process called forest gardening. Forest gardens originated along rainforest river banks and in monsoon-swept foothills. Initially useful plant species were identified, protected and improved, while undesirable species were eliminated. Later on, other desirable foreign species were introduced into the gardens. Initially the hunter-gathering way of life continued alongside forest gardening, but over time, due to the pressures of population from encroachment or increased competition leading to a decline in the availability of wild foods, particularly game, a pastoral lifestyle took over.

The first agricultural revolution took place during the Neolithic period around 10,000 years ago. This gentle progression from nomadic foraging to forest gardening became a widespread transition to agriculture and settlement. Archaeological evidence shows that various forms of plant cultivation and animal domestication evolved independently of each other in at least six separate locations around the world. This move was not just a change of lifestyle, but a complete change in the process of energy management; from the procurement of energy directly from plants and animals by hunting and foraging to the storage of energy derived from soil fertility in the bodies of domesticated stock and grasses. This process of development was rapid, especially in Europe and the Middle East, and it led to a significant change in management both of the land and of labour organisation, tying people to a single place for the first time.

Agriculture did not bring an end to hunting per se, and we should draw a distinction between the sedentary husbandry peculiar to farms and the relatively nomadic existence of herders, but the seasonal migration of big game made it sensible to keep animals close at hand, to the point of actively bringing them into the settlements themselves. The size, diet, temperament, behaviour and lifespan of animals were all determining factors in their suitability for domestication.

A 50-year experiment by Russian scientists into the domestication of silver foxes has given us an insight into how our Neolithic forebears might have achieved this. The breeding programme was set up in 1959 by a Soviet scientist called

Dmitri Belyaev who was interested in the process by which, 15,000 years ago, wild wolves became tame domestic dogs. Cats and dogs fill the same evolutionary niche, and evidence suggests they have similar intellectual capabilities, yet it was dogs that became man's best friend. Belyaev believed the key factor in the domestication of dogs was not size or strength, but behaviour – they are simply friendlier than cats.[3] In his experiment, he visited Siberian fur farms and selected animals for breeding, choosing only those that were not aggressive and allowed researchers to come up closest before running away. He believed that this selection criterion for 'low flight' would mirror the natural selection process. The results of the breeding programme showed that within just a few generations the animals were passive enough to be handled, and after just five generations were completely tame. By the eighth generation, the animals were totally different in temperament and behaviour from their 'wild' cousins, also displaying changes in physiology – different fur patterns caused by lower adrenaline production. In short, they had become much more 'dog-like'.

Dogs, as we know, can be enormously helpful animals. They can be used to hunt or to herd, and they can be used to provide protection or to control vermin. Other domesticated species too were chosen for their usefulness. Goats and sheep are the basic subsistence animals. They require very little specialist knowledge to look after, are easy to feed, providing a renewable source of protein (milk), wool and fur during their lives and meat and skins after they have died. Cattle present more of a challenge, but are also much more useful. As well as food and skins, cows also provide fertiliser and can be put to work ploughing a field or towing a sled.

Selective breeding is a process almost as old as farming itself. The eight 'pioneer crops' of the Neolithic period – emmer and einkorn wheat, barley, lentil, pea, chickpea, bitter vetch and flax – were not just bred for greater calorific

---

[3] Or if you are a cat-owner, 'less independent'.

returns but for qualities such as flavour and seed retention. Other pioneer crops were tried and abandoned, then taken up again at a later date. Rye, for example, was originally grown in Anatolia, then discarded before being successfully domesticated several thousand years later in western Europe. Other plants presented a different challenge. For example, wild lentils are biennial, but ancient farmers managed to break this dormancy by selective breeding and develop a strain that would produce an annual crop.

These steady improvements in agricultural processes meant that for the first time ever there was an extensive surplus of food. The nomadic lifestyle of hunter-gathering made food storage impossible, but early settlers did not have the same problem. Granaries and smoke houses were developed and they allowed food stocks to be preserved over a much longer term. The availability of food led to population expansion and over millennia transformed human society from mobile to sedentary. These settlements became larger, allowing communities to develop specialist workers and more advanced tools. Society became much more like the one we know today, with occupations and niches. This labour diversification led in turn to the formation of higher-density villages and towns and the first trading economies in which ideas, as well as goods and services, could be exchanged.

None of this should lead us to assume that life on a Neolithic farm was idyllic, however. Far from it. We should view this agricultural revolution as a necessary change rather than as a move forwards. It's true that agriculture made it possible to provide a subsistence existence for a much greater population, but the cost was significant. For many, farming was a necessary evil, nothing more; an effective means of providing food for the table, but one that ensured a hard life of toil and semi-misery in return. Farming was a much harder way of life than hunter-gathering and much more time-consuming too. Farming didn't make people happier, and the quality of life diminishes almost as soon as farming starts. The size of hunter-gatherers' ranges shrank because of pressures of population, so farming was a necessary strategy for survival. The equation it solved is that of calories per hour foraging versus farming. A successful large beast hunt yields many thousands of calories but needs a big range to ensure repeated successes. Furthermore, the diet that farming provided was less nutritious and diverse. The nutritional standards of Neolithic people were inferior to those of their hunter-gatherer forebears. This is reflected in the average height of the adult population, which fell from 178cm for men and 168cm for women to 165cm and 155cm respectively. The poor quality of the average diet kept average heights below pre-Neolithic levels until the 12th century.

Agriculture was a compromise, not a step forward: an effective solution to the challenges caused by population growth and increased competition for resources, but not one that provided a better, easier or healthier lifestyle. And we should bear in mind that in parts of the world where competition for resources was not so fierce – in North America, Australia, Africa and parts of north and southern Asia – the hunter-gatherer life prevailed.

The shift to organised food production supported a denser population, which allowed larger sedentary communities to flourish. To operate effectively, these bigger societies required decision-making and organisation, which paved the way for government and the social elites; the first steps down a path that led to soldiers, priests, bureaucrats, politicians and celebrities. Food surpluses meant that these social elites could focus on the process of governing without having to engage in food production, maintenance or commerce, which allowed them to dominate decision-making and paved the way for modern societies organised by laws and ruled by governments. The agricultural revolution of the Neolithic period can been seen as the first instance of a response to a population–food–energy crisis. It would not be the last.

5 States of Emergency

'Civilisation is the limitless multiplication of
unnecessary necessities.'
Mark Twain

It's tempting to think of a population–food–energy crisis as
a problem exclusive to the 21st century, yet the challenges
we face today are as old as human civilisation itself; and those
associated with climate change are even older. Since the
first settlements became organised, regardless of who was in
charge, the salient question has remained: How do we feed
everyone and keep them safe and warm? Not least because
a cold and hungry population is likely to cause a lot more
trouble than one that's well fed and snoozing contentedly in
their beds. Throughout history, successive civilisations have
faced challenges on the same population–food–energy–climate theme. In some cases, like the Neolithic revolution,
society has adapted its lifestyle to overcome them successfully, but there are many examples where the outcome has
not been so rosy.

The culmination of the Neolithic revolution can be seen
in Sumerian cities, built of mud, which flourished on the
banks of the Tigris and Euphrates in the Middle East some
5,500 years ago. The Sumer people settled in Mesopotamia
(modern-day Iraq) between 4500 and 4000 BCE, establish-
ing a dozen or so independent city states – each containing
several thousand people at their peak – which lasted for over
2,000 years. The Sumer appear to have invented the wheel,
as well as cuneiform: if not the first, then certainly one of the
earliest examples of written language. It was also the first civilisation to experience a collapse. As the number of citizens in
the cities increased, pressure was put on the land to produce
more and more food. This over-farming caused an increase in
soil salinity and a calamitous reduction in productivity. There

were palliative attempts to cope with this environmental disaster. Wheat cultivation ceased and was replaced by barley, a more salt-tolerant crop, but it was to no avail. Over time, fertile arable land became barren desert and the Sumerian population fell by over 60 per cent.

The Sumer lacked the knowledge to cope with their change in circumstances; they were unable to relate their problems to either causes or solutions. Yet the process of increasing urbanisation they began has continued ever since, punctuated with collapses for much the same reasons: population growth, famine, energy shortage, environmental change or any combination thereof. Although the threats have been perennial, the outcome is never so straightforward. There have been times when the impact has been sudden and catastrophic; others with a slower but equally terminal decline; and yet more still where the outcome has been a return to growth and prosperity. If crises are inevitable, then there are lessons to be learned from the past. Certainly, it appears that overcoming these challenges requires a combination of ingenuity, application and good fortune in equal measure.

At the moment the Sumerian society was imploding, on the other side of the world in Central America an even more mighty civilisation was beginning to establish itself. The Maya flourished from 2000 BCE to 900 CE, extending from present-day southern Mexico to parts of northern El Salvador and western Honduras. The Maya are notable for having the only fully developed written language in the pre-Columbian Americas, which allowed them to develop the sophisticated system of sustainable government (and taxation) required to run an empire for any length of time. At its peak, between 250 and 900 CE, several Mayan cities boasted populations of tens of thousands, with the capital Tikal containing an estimated 100,000 souls.

This achievement is all the more impressive when one considers that the Maya lacked the access to heavy draught animals – horses, oxen, camels, buffalo, elephants – enjoyed by their Eurasian and African counterparts. They had to rely on human muscle for the heavy work. Despite or perhaps because of this, they developed diverse and sophisticated methods of food production. Most agriculture took place on permanent raised fields, arranged in terraces, but highly advanced forest gardens and managed fallows, as well as organised wild harvesting, were also crucial for supporting the large urbanised population of the civilisation's classical era. Contrary to popular belief, the Maya did not vanish mysteriously into the ether, or under the ocean; their direct descendants are alive and living among us today. But their complex way of life did disappear. By 800 CE their cities were abandoned, their technological expertise forgotten, and the few people who remained had returned to a simpler life. There is not one universally agreed theory as to why the Mayan civilisation declined, but the most popular hypothesis is that it was the result of over-population and climate change. There is evidence that the Maya exceeded the carrying capacity of their environment, exhausted the land's agricultural potential and over-hunted the region's indigenous megafauna to extinction. The most recent research, using high-resolution climate models and new reconstructions of ancient landscapes, suggests that converting much of their forest land into cropland may have led to reduced evapotranspiration and, in turn, of rainfall. The natural effects of drought were greatly magnified, so much so that one appears to have lasted for almost 200 years.

The Maya's challenges quickly turned into catastrophe, but in Europe the fall of the Western Roman Empire was even more precipitous. For almost 800 years Rome was the dominant power in western and southern Europe, North Africa and the Middle East; its empire stretched from the Tyne to the Euphrates. The Romans faced many problems that would be familiar to us today. The population density in their empire's eponymous capital, a city of 1 million people, was greater than 21st-century New York City. But for most of its history, the Roman empire was as it appeared: a vast, technologically advanced and seemingly unassailable state. Yet it disappeared over the space of a few decades, plunging Europe into the original Dark Age. As with the Maya, there was no single reason why it failed, but population growth was a major contributing factor. There were pressures across Rome's borders with Germany and eastern Europe, due to the various Germanic and Slavic tribes increasing competition for resources. What began as a steady trickle of refugees became a series of full-blown invasions.

Notably, neither the Maya nor the Romans had the wherewithal to develop any technological solutions to their different challenges. These empires were big, but they were also inefficient. The Mayan cities, huge and impressive though they are, were rather slapdash in construction, town planning determined more by typographic expediency than demographic effectiveness. The Roman empire in particular had no culture of innovation at all – not just at the end, but across its whole 800-year history. In fact the Romans invented little of any significance themselves, preferring to import technology in much the same way they imported religions or legionaries.

The Romans did have a formal education system, but they were no proto-Montessoris. There was no free education, but children of middle- and upper-class families were taught in what were literally grammar schools. A Roman education consisted essentially of learning hundreds of classical texts by heart, and then writing them out again and again and again. Encouragement in this task consisted of levels of corporal punishment that even the Chilean secret police might have considered excessive. The mindless repetition was relentless – there were no weekends and few holidays, the boredom broken only by the occasional thrashing. This created a culture in which innovation was a cultural taboo, and these moribund, monotonous and cruel years of primary education left the empire's future ruling classes naturally incurious, dismissive of learning, with no respect for either knowledge or its application. The greatest and brightest minds were coached systematically to be woolly and mediocre.

For the implications of this, one need only look at what the Romans could have invented, but didn't. Perhaps most spectacularly, despite having ink and paper (or papyrus), despite knowing how to stamp complex designs onto coins and other metals, and despite an abundance of olive presses that could have been adapted specifically for the purpose in a matter of days, the Romans did not invent the printing press – or indeed printing of any kind. Books were mass-produced by scribes, usually literate slaves who worked in their hundreds, copying out texts in longhand. Not one person ever thought that there might be a more effective method to do this. We know what a difference printing would have made, because of what happened in Europe during the Renaissance.

In the first century CE, Heron of Alexandria created the world's first steam engine. It was called the aeolipile. Its design was pretty basic, but it seems to have made it to the prototype stage – for a time it was even used to open the doors of a temple, no doubt to the wonderment of the congregation. And there things ended: nobody could see the possibilities. Even though the Romans had railway tracks, trucks, access to coal and petroleum, and the ability to make high-quality steel, the aeolipile was viewed as little more than a curio. The industrial revolution could have been a 2nd- rather than a 19th-century phenomenon, but, the battlefield aside, there was no progress at all. If a Roman citizen born at the beginning of the Pax Romana of Augustus were to travel through time to the reign of Constantine I, some 350 years later, they would find that very little had changed. By way of comparison, that would be like someone from the reign of Charles II visiting us today.

In Central America civilisation did not return to technological levels achieved by the Maya until after the Columbian discovery. In Europe it was almost 800 years until comparable levels of technology were achieved, and 1,000 years before cities could rival the size and grandeur of the empire's capitals.

After the fall of Rome, things unravelled pretty quickly in Europe. The Dark Ages mark a period of about 200–300 years immediately after the fall of the Roman empire. When they start really depends upon where you are. Although 476 CE is traditionally regarded as the year the empire fell, it didn't disappear overnight. In Britain, for example, the Dark Ages mark the period between about 400 CE – when the Roman legions began to leave and southern Britain was dominated by a Latin-speaking, Christian, Romanised landowning class – to about 600 CE when the Anglo-Saxons started to write things down and history started again.

In 400 CE the population in Britain was relatively urbanised, living in tenements made of brick, stone, concrete and tile, under a centralised government that raised taxes that were paid using minted coins. This revenue was spent on roads, public buildings and, most importantly, a vast standing army. In the countryside, the ruling classes lived in centrally heated villas and dedicated their spare time to the nuances of Latin grammar and writing poems to their friends. Goods were manufactured and sold commercially and there was trade across an empire that stretched from Hadrian's Wall to Mesopotamia. There were schools, public baths and economic production. When history begins again in 600 CE we find a very different picture. This Romanised class has disappeared, gone too are their tenements and villas, while economic production has virtually ceased. Most goods are now produced at home rather than bought commercially. The population has massively declined, coins are no longer used for exchange, towns have disappeared and brick and stone buildings have been replaced with wattle-and-daub huts. There are virtually no buildings above two storeys high. The old imperial provinces have been replaced by 20 or more small kingdoms that owe nothing to the political geography of the empire. We might not know how it happened but we do know what happened: typical Romano-British lifestyles rapidly disappeared as soon as the legions left.

The population of Europe remained fairly static until

the mid to late Middle Ages. But by 1250 the economy was benefiting from a rapid increase in the population, with the number of people reaching a peak that would not be seen again until the 19th century. This particular growth spurt was checked by a series of events in the 14th and 15th centuries known collectively as the crisis of the late Middle Ages, which brought growth and prosperity to a crashing halt.

By the latter decades of the 13th century Europe had become over-populated. Frontiers had stopped expanding and internal colonisation was coming to an end. The population was not just high, but also young. At the beginning of the 14th century, 80 per cent of people living in England were under 20 years of age; experience and expertise were in short supply. With the people already living at the edge of natural resources, climate change brought an end to the medieval warm period, ushering in a 'mini ice age' of harsh winters and smaller harvests. Across northern Europe, the land became harder to till, innovations such as the heavy plough and the three-field system proving less effective than in Mediterranean countries, due to harder, less loamy clay soils. Famine, under-employment and inflation became endemic; scarcity was the natural order of things. By the start of the 14th century malnutrition was rife, which in turn reduced immunity to disease and raised mortality rates. In the autumn of 1314 extraordinarily heavy rainfalls resulted in a catastrophic famine, the worst in European history, which reduced the population by more than 10 per cent.

Governments attempted to provide an economic response by prohibiting the export of foodstuffs, fixing the price of cereals and other staples. They also outlawed large-scale fishing, hunting and foraging to conserve supplies. These measures proved at best ineffective and in most cases actually exacerbated the problems. The countries that were hit hardest, including England, were unable to import any grain, and most of the little that was shipped in found its way onto the flourishing black market.

With the European economy in the midst of a downward

spiral involving chronic hunger, low productivity and rising international conflict, its already debilitated citizens succumbed to a series of virulent epidemics. In the opening decades of the 14th century many thousands across the continent died of typhoid, with those living in urban areas the worst affected. In 1318 an unidentified disease, probably anthrax, wiped out livestock populations, further reducing food supplies and affecting incomes of everyone involved in husbandry, from landowner to peasant.

Yet all of these hardships, great though they were, proved to be merely a warm-up to the main event. In October 1347, twelve Genoese galleys pulled into the harbour at Palermo in Sicily and in doing so introduced bubonic plague into Europe. There were almost certainly subsequent introductions, but the initial Sicilian outbreak is the easiest to trace. Within two weeks the Black Death had reached Pisa and mainland Italy. The local response was to expel the galleys, but this only succeeded in hastening the spread of a virulent disease for which maritime travel was to prove the evolutionary jackpot. In January 1348 the plague had arrived in France, and by the summer it was spreading rapidly across Spain, Portugal, England and Germany. Within infected areas it killed, on average, between 45 and 50 per cent of the population. However, not all places suffered equally. The devastation was worst in Spain and France, where plague ran for four years consecutively and the death rate was between 75 and 80 per cent. In countries where fewer people lived in towns and cities, such as Germany and England, death rates were closer to 20 per cent. Whatever the breakdown, the impact on human numbers was enormous; half of Paris's 100,000 residents perished and by 1400 plague had accounted for the lives of between 75 million and 200 million Europeans.

The Black Death affected the lives of sufferers and survivors alike. Those who lived through the plague found that the shape of society had changed completely. In England, for the first time, labour was at a premium. Peasants and agricultural workers were no longer tied to a landowner. The competition for their labour meant that they were able to work for whoever paid the best price. The change in economic circumstances created a mobile, transient population and parts of the country became depopulated. In Cumberland and Westmorland, estates fell into ruin simply because there was no one to work the land. The changes were sociological as well as geographical. The population became progressively more urban. By 1550 many towns and cities had regrown in size to pre-plague levels, despite the fact that the overall populations remained much lower. Prior to 1350 there were about 170,000 settlements in Germany. This number fell to less than 130,000 by 1450, but the concentration of people living in towns and cities was much greater. In 1350, London and Paris had populations of 50,000 and 215,000 respectively, which grew to 200,000 and 325,000 by 1600.

By the middle of the 16th century the population had stabilised and food supplies were keeping relative pace with demand, but Europe was faced with yet another crisis, this time one of energy. The Elizabethan energy crisis (which ran from 1570 to 1630) has a remarkable number of similarities to our own position today; as we contemplate peak oil, so the Elizabethans were coming to terms with peak wood and the startling revelation that their exploitation of what had been regarded as a cheap and plentiful supply of fuel was completely unsustainable. A man embarking on a career as a woodcutter in the 16th century could expect to fell around 30,000 oak trees during his working life; each of those trees would have taken 80 years to grow. Wood was the most important natural resource for the Elizabethans, providing not just domestic heating but one of the fundamental materials required for construction, shipbuilding and, through the production of charcoal, for the iron and other metal-smelting industries as well. The highest demand for good-quality oak came from Britain's vast naval and merchant fleets. Around 650 fully grown trees were required to produce just one galley. Initially few people gave a thought to deforestation, indeed many landowners welcomed the process as it cleared land that could be put to more profitable use such as grazing for sheep (in 1650, around 85 per cent of UK exports were provided by the wool trade). However, people soon began to realise that rampant tree-felling was laying waste to vast swathes of the countryside. In parts of Britain not a single tree could be found standing for many miles around the larger population centres. Firewood was becoming expensive and increasingly difficult to get hold of. The unsustainable consumption of timber would have undoubtedly had a catastrophic impact on society, perhaps even leading to a Mayan-style collapse, had it not been for the fact that there was a readily available, under-used alternative fuel close at hand.

Coal was a product of no great importance prior to the 16th century. Although the Romans and Greeks were aware of coal's energy properties, its use was limited. In Britain during the Middle Ages it was known as seacoal, because much of it was found in coastal areas where seams are often exposed to the surface. Coal was transportable by sea and could generate tremendous heat for industrial processes. It is also a far more effective fuel than wood. The energy density of coal is much higher, which means that pound for pound the same medieval transportation system was able to move much greater supplies of energy to urban areas without any additional infrastructure or distribution costs. Getting at the coal itself was a different matter. In 1260, coal was being shipped to London from Newcastle for use by blacksmiths and lime-smelters, but by the time of Elizabeth I all the easily accessible sources had been exhausted and the process of mining underground, which had begun in the 13th century, became widespread.

The similarities between the Elizabethan energy crisis and our own are notable. While wood remained plentiful there was little incentive to mine coal. However, as practical resource limits were approached, the use of coal became accepted. When wood became scarce near towns and cities there was a strong motivation to shift to energy-dense coal instead, much the same as the transition from coal to nuclear or solar that many propose today. Secondly, the scarcity of wood proved the trigger for a shift to a more sustainable and cheaper form of energy (albeit only for a few hundred years). The adoption of coal was not without its detractors. Coal was seen as dirty and polluting, a fuel of last resort, while deep mining was regarded as a form of robbery from the Earth until the early 16th century, which echoes environmental sentiment today.

The Elizabethan energy crisis marks the point at which we officially disconnected from the solar cycle and began consuming more energy than we were able to renew. Once our love affair began with these stores of sunlight formed over many millions of years, we never looked back. The growth in the use of coal, both during and after the Elizabethan era, led to innovations in materials to deal with the potent heat produced. Houses were built of brick and stone rather than timber and plastered willow. Windows, previously covered with waxed hessian or paper, were now glazed; the glass itself a product of coal-fired furnaces. Chimneys allowed rooms to be heated independently, made multiple-storey buildings possible, and also removed the fumes and smoke, improving the air quality indoors at least, if nowhere else. The early modern world quickly learned to enjoy the liberating and comforting domestic benefits of coal, but it was in the world of commerce, through the developments of the industrial revolution, where its impact would be felt most. In 1700, 80 per cent of the world's coal was being produced by Britain, and once the steam engine took over from waterwheel and windmill, the period of solar deficit was well and truly under way.

6   The Solar Deficit

'Humans are about 10,000 times more common
than they should be.'
*Steve Jones, Professor of Genetics, University College
London*

'Can you please quit your rutting for just a couple of
seconds while we figure out this food/air deal?'
Bill Hicks, comedian and satirist

The Elizabethans came through their energy crisis not just
with a replacement fuel for wood, but with a raft of opportunities that once firmly grasped were to transform an
inconsequential European backwater into the world's foremost industrial power. At the time of Elizabeth's accession
England was reaching the nadir of decades of decline. The
dotage of Henry VIII had given way to the political and religious uncertainty of the child king Edward IV, whose brief
reign was marked by economic problems and social unrest
that erupted into full-blown revolt and rebellion. Edward
was eventually succeeded by his eldest sister Mary, but those
hoping for a return to the good old days were to be disappointed as poor harvests, famine, bitter winters, foreign
policy failures, religious persecution and widespread public
discontent marked a dismal five-year reign.

    The heart of the problem was an endemic shortage of food
and energy. The shift from a wood-burning to a coal-burning
economy went a long way towards addressing these issues.
Coal not only provided a means of keeping people warm in
winter, but along with the process of enclosure (a systematic
ending of the right of common access to village pastures and
meadows) it was largely responsible for driving the industrial
revolution. Enclosure provided the environment for more
efficient and intensive farming, which increased agricultural
output but led to widespread unemployment among the now

redundant rural population. By fuelling industrialisation, coal helped to create a new market for this displaced labour in burgeoning mines and factories. The economic success of towns and cities that followed led ultimately to our more secular, predominantly urban way of life. By disconnecting from the solar cycle – where the total amount of sustainable energy was equal only to the amount of wood produced by a year of photosynthesis – a community living on the very edge of its means found the resources to rapidly expand. Where England led, the rest of the world followed. The coal-fuelled expansion was indeed rapid. In 1658, the year of Elizabeth's accession, the global population was around 500 million. In just 200 years this increased to 791 million; an increase of over 50 per cent.

This exponential growth did not go unnoticed, and several eminent 18th-century political thinkers speculated as to whether the increasing number of people could be sustained indefinitely. The prevailing view was that it could, and nobody embodied this optimism better than William Godwin. Godwin was a typical Enlightenment all-rounder: a gentleman, essayist and provocative leader-writer who was intrigued by the scientific breakthroughs of the age. He was also fond of the odd thought-experiment, particularly in respect of human nature and how it might be linked to population growth. Godwin was a Utilitarian pioneer and a great influence on Jeremy Bentham and John Stuart Mill, but he pushed the notion of 'the greatest happiness of the greatest number' to the extreme with his belief in the principle of 'the perfectibility of society'. In essence, Godwin contended that as societies became more affluent they would also become more advanced intellectually, so people would be able and willing – by principles of pure reason – to do whatever was best for the community. As such, while population growth is simply the natural way of things, once the number of people reaches a certain level, all good citizens will do the decent thing, which is to consume less and stop having kids. Godwin's focus on self-government is the reason why he is regarded as the first modern proponent of anarchism, but although his rather rose-tinted view of human nature resonated widely, it was not shared by everyone.

In 1798, Thomas Malthus wrote his first *Essay on the Principle of Population* in direct response to William Godwin and other similar-minded writers, notably the Marquis de Condorcet. Malthus was a gentle and personable man with a wide circle of friends. He was erudite, cultured, entertaining and by all accounts hugely likeable. He was father to no fewer than eleven rosy-cheeked daughters. The main tenets of his argument were in radical opposition to the popular thinking of the time. Malthus noted that contrary to what Godwin proposed, populations were inclined to increase during times of plenty and that the only thing that appeared to stem this growth was strife, whether through famine, disease, war or some other natural disaster. Populations, he said, were 'doomed to grow' until strife occurred – usually afflicting the poorer members of society the most – which diminished the ability of the world to feed itself:

> Let us imagine for a moment Mr Godwin's beautiful system of equality realised in its utmost purity, and see how soon this difficulty might be expected to press under so perfect a form of society ... Let us suppose all the causes of misery and vice in this island removed.
>
> War and contention cease. Unwholesome trades and manufactories do not exist. Crowds no longer collect together in great and pestilent cities ... Every house is clean, airy, sufficiently roomy, and in a healthy situation ... And the necessary labours of agriculture are shared amicably among all.
>
> The number of persons, and the produce of the island, we suppose to be the same as at present. The spirit of benevolence, guided by impartial justice, will divide this produce among all the members of the society according to their wants ... With these

extraordinary encouragements to population, and every cause of depopulation, as we have supposed, removed, the numbers would necessarily increase faster than in any society that has ever yet been known ...

Malthus recognised that availability of resources determines population growth, but he differed significantly from Godwin in that he felt over-population – where the number of people exceeds the carrying capacity of the environment – and not equilibrium was the most likely outcome.

Long after his death, many of Malthus' principles regarding exponential growth were confirmed in animal populations. The underlying problem that Malthus identifies is that output increases linearly whereas population grows geometrically and even exponentially. In micro-organisms, exponential growth occurs until an essential nutrient is exhausted. Think of a single bacterium on a layer of jelly in a petri dish. Typically this will split into two daughter organisms, which will in turn split into four and so on until the food supply in the dish is exhausted, at which point the population will collapse. Transferring the culture to a bigger dish will lead to a bigger colony, but once food is exhausted it will collapse in just the same fashion. Human societies are much more sophisticated than bacteria and their energy requirement takes many different forms, but Malthus argued that their population would continue similarly to expand whenever there were resources available, going way beyond the point at which such an expansion was sustainable – and that once those resources were exhausted, it would also collapse.

Malthus challenged several widely-held notions – like the idea that increased fertility was an economic benefit because it provided industry with more workers – and proposed some radical solutions to the problem, such as freely available contraception. Unsurprisingly, then, his views proved to be widely unpopular and the genial Reverend became the target for a great deal of vitriolic personal criticism. The poet

Percy Bysshe Shelley called him a 'eunuch and a tyrant' while others questioned whether a man with eleven children was in any position to be doling out lectures on population control. His views were also misconstrued by many as an attack on the poor rather than on poverty. Marx called him 'the principal enemy of the people' while Dickens parodied Malthusian support among the wealthy in the character of Ebenezer Scrooge, who press-ganged one of his arguments to excuse his own parsimony when it came to charitable donations: 'If they would rather die they had better do it, and decrease the surplus population.' By the late 19th century Malthus' ideas had completely fallen out of favour, most economists believing that improvements in technology, science and finance rendered all of his arguments implausible.

Yet much of what Malthus predicted has indeed come to pass. His theory of supply–demand mismatches, which he called gluts, was dismissed as ridiculous on publication, but over 100 years later it was clear he had foreshadowed one of the major causes of the Great Depression, which resonated strongly with John Maynard Keynes. Malthus remains a strong, if not incontrovertible, influence on population science and economics today. And not without good reason. The body of evidence available to us some 200 years later is more sophisticated than that available to Malthus, working half a century before the publication of *On the Origin of Species*, but many of his ideas still seem relevant. Human population does indeed seem to increase during times of plenty until it inevitably overreaches itself. However, while collapses do occur – as experienced by the Sumer, Maya and Romans and many others – thanks to ingenuity and invention they are not inevitable.

※

For *Homo sapiens* the first 'bacterium moment' took place around 70,000 years ago, when a supervolcanic eruption occurred on the modern-day site of Lake Toba, Indonesia. The Toba event is accepted as one of the largest eruptions of all time. It plunged the planet into a decade-long volcanic winter and subsequent 1,000-year ice age. This radical climate change reduced the global human population to little more than 1,000 breeding pairs. Today, geneticists can trace the entire human race – all 7 billion of us – to a single male living at that time (known to his friends as Y-Chromosomal Adam).

The Toba event was a natural disaster but it was not a catastrophe for the human survivors, who seized the opportunity presented by a dearth of competition and enjoyed a period of rapid population growth, innovation, progress and migration. By the time of the Neolithic revolution, the global population had risen to 5 million, which in many parts of the world put it beyond the limits of sustainability for a hunter-gathering lifestyle. The adoption of farming as a way of life increased the carrying capacity of the environment massively, with between 200 and 300 million people sustained by 1000 BCE.

It's important to remember that the world population was in no sense 'globalised' at this point. All the major land masses had human inhabitants by 10,000 BCE, but they were not all connected and the population grew at different rates in different parts of the world. In Eurasia the population continued to rise steadily, due largely to improvements in farming and to more widespread distribution of surpluses as a result of trade, until 541 CE when the 'Plague of Justinian' – most likely an early strain of bubonic plague – devastated the Eastern Roman Empire, halving the population of Europe in the process. However, in Australia, a similar-sized continental land mass, the number grew very slowly to just 750,000 by the time of its discovery by Europeans. Despite the discrepancy, across the globe we find this same pattern of steady growth during times of plenty, disrupted to varying extents by plague, warfare, famine or climate change. By 1340, the world population had grown to approximately 450 million people, with 70 million of those living in Europe. This is probably about the maximum number of people that is possible given a medieval technology level, diet and life expectancy without relying upon fossil fuels. It's certainly reasonable to argue that by killing 100 million people in the 14th century, the Black Death postponed the Elizabethan energy crisis by 300 years, which is the time it took for the global population to recover.

The adoption of coal marks the point at which we escape from our earthly petri dish. Its relative abundance, along with that of gas and oil, has extended our ability to exceed the carrying capacity of the environment for almost 400 years. Coal's impact on population growth lay in the fact that it was more than just a replacement fuel for wood. Stocks of coal were so abundant that they could be regarded as inexhaustible. Coal burns more slowly than wood, at a higher temperature, and provides more heat, but it also produces more smoke, which meant that fires needed to be contained in brick or stone chimneys. Larger rooms and bigger buildings could be heated more efficiently, and fortunately coal also provided a means of mass-producing the bricks that these new constructions would require in a much more efficient and cost-effective manner.

Coal also contributed to a change in the economics of energy: for the first time, fuel became fully commoditised. Plentiful it may have been, but it was plentiful deep underground. Unlike firewood, which people could gather for themselves, coal could not be mined by individuals. Its extraction required organisation and large numbers of labourers: if you wanted coal, you had to buy it. Over the next 300 years there would be a shift away from a largely rural, peasant population eking out a subsistence living directly from the land using a system of barter as the basis of trade, to an urbanised working class who were paid a subsistence wage in return for their labour. Subsidising one's basic wage with foraged resources became impossible for most people; everything had to be paid for: shelter, food, fuel and clothing.

Coal provided a fuel for the industrial revolution that followed, but it wasn't enough on its own. Industrialisation requires metal. Take a look around and you will see that we live in a world constructed almost entirely of metal: buildings, roads, railways, cars, kitchens, plumbing, telecommunications, computers … even the things we use that aren't made of the stuff, almost certainly relied on it at some point in their production. We take metal for granted, but its relative rarity in the late medieval world is one of the things that, were we able to visit, we would find most striking. Humans have been working metal since copper was first smelted from ore by Neolithic tribes living in the Middle East some 7,000 years ago. These copper tools had huge advantages over the flint and bone objects they replaced. Copper knives can be resharpened to retain their cutting edge, and unlike stone, metal is malleable, so it can be worked into a variety of different shapes. Even when metal blades and tools break, they are relatively easy to recycle. Yet it's one of the more curious properties of metal that makes it so useful: the more you hit it, the harder it gets.

The ancient smiths knew this, but it's only within the last 100 years that we have really understood why it happens. The atoms in metal are arranged in a lattice of neat ranks and files, which are quite easy to move around. When the metal is beaten, the atoms become bunched closer together in little knots, within which there is very little movement, and it's these knots that give worked metal its strength.

Combining different metals into alloys can produce surprising results. The Sumerians began adding tin to copper around 2900 BCE. The resulting alloy, bronze, was much stronger and more durable than its soft components and it could be worked into a variety of tools and weapons. Each block in the great pyramids was fashioned by hand using chisels made of bronze. Iron, one of the commonest elements in the Earth's crust, is even harder than bronze and more durable still, but working it is extremely difficult. What emerged from the early furnaces was a glowing spongy lump called a bloom (hence the name of these workshops, 'bloomeries'). The blacksmith would have to consolidate these blooms by hammering out the impurities (the sparks that fly off during the beating), leaving a small iron bar that can then be worked – or wrought – into something useful at a forge.[1]

The problem is that it's not possible to get the temperature of a wood fire hot enough to smelt iron, so a different fuel – charcoal – had to be used. Charcoal is wood with all the water and other impurities removed, and so it has a much higher carbon content. In many ways it's similar to coal, and it burns at a greater temperature than wood, but even with charcoal it wasn't possible to build furnaces that were hot enough to melt iron until the 13th century. Despite these advances, by the late Middle Ages mass-producing iron items, like nails or cannon shot, was still expensive: a labour-intensive process requiring huge amounts of wood to be turned into charcoal. In the 16th century, the demand for iron goods increased at the same time as deforestation was affecting the supply of charcoal to smelt. Coal again provided the solution, with coke (its partially burned derivative) proving to be an excellent substitute. While the impurities in coke proved problematic in the traditional wrought-iron smelting process, the new coke-fuelled blast furnaces were hot enough to melt iron, allowing it to be poured into moulds instead. This 'cast iron' was invented by Shadrach Fox, ironmaster at Coalbrookdale forge in Shropshire, who was casting grenade shells for the Board of Ordnance in 1693, but credit is usually given to his successor Abraham Darby who, if nothing else, certainly succeeded in making the process commercially viable.

---

[1] This process of transforming iron ore into the base metal relies on reduction, which in turn involves the oxidation of carbon – by burning coke, coal or charcoal – to provide the energy. Oxidation and reduction are the chemical processes that led to the creation of the modern world. Photosynthesis reduces carbon dioxide to carbohydrate, releasing oxygen. The oxygen combines with iron to form iron ore, which refining reduces to make base metal. All stages are powered by oxidising carbon and carbohydrate to make carbon dioxide. It's the sun that powers all these transformations – although sometimes they are separated by billions of years.

Fox and Derby's casting process reduced the costs of mass production and created an opportunity to exploit Britain's vast quantities of iron ore. The 18th century was the metal's golden age. Much like plastic today, anything that could be made out of iron, was: bridges, ships, steam engines, packaging, cutlery, artificial limbs, even false teeth (US President George Washington owned at least two sets of iron dentures, which must have been remarkably uncomfortable to wear). Most importantly, cast iron could be used to make machine components, which automated processes that had previously been carried out by hand. These innovations – Crompton's Mule, Arkwright's Water Frame, Kay's Flying Shuttle et al. – have been a staple of economics and social history lessons ever since.

As a material, even iron was not without its drawbacks. It's a highly reactive metal, which makes it prone to corrosion, while impurities that cannot be removed during the casting process make it structurally suspect. Iron is prone to buckling at relatively low temperatures and snapping when stretched and compressed repeatedly. Steel, a mixture of iron and carbon, is a much better material. Steel is stronger, harder and lighter than iron and is much more resilient when either stretched or compressed. Mass production of steel finally became viable in 1855 as a result of the process patented by Henry Bessemer, which reduced production costs from £40 per ton to just £7 per ton; making it as cheap to produce and purchase as wrought iron. With this cheap steel we were able to create the modern world. Steel was used to construct the first skyscraper, the Home Insurance Building in Chicago in 1884. It's due to steel that buildings became taller and taller. Beyond construction, the development of steel cable, steel rod and sheet steel enabled large, high-pressure boilers, and high-tensile-strength steel for machinery made it possible to create more powerful engines, gears and axles than ever before. Industrial steel also allowed the construction of giant turbines and generators that could harness the power of water and steam.

By the end of the 18th century coal was in such widespread use that it had become the dominant energy vector. The economy was now completely unfettered from the solar cycle, and the opportunity to create a very different world was seized with blackened hands. Rapid mass transportation, initially in the form of locomotives running on a burgeoning rail network, became possible for the first time, thanks to coal's massive energy density and portability. Coal's versatility meant that it could be easily adapted to a range of purposes from heating homes to powering factories; it reduced the price of production for everything from steel to shipping. Not that coal improved the lives of everyone it touched, or even the majority. The global population reached 1 billion people in 1804. In the towns and cities of Europe and the USA, most of these people were workers living in high concentrations and squalid conditions, near to the factories or mines within which they would spend almost every daylight hour of their lives. The personification of hard labour is surely the engine-room stoker: furiously feeding the ship's furnaces when the captain calls for full steam ahead. And so it was for miners and factory workers across the industrialising world. For them, the introduction of coal meant brief lives of toil, and in many cases, abject misery.

The economic impact of coal is clear to see, but the final revolutionary use for the fuel was arguably its most profound. Coal in its natural state is costly to transport. It requires vast distribution networks to service domestic and industrial customers, often hundreds, if not thousands of miles away from the mines themselves. Where there is expense and inefficiency, and a dollar to turn, enterprise will often find a way. Nobody wanted coal, what they wanted was the energy that coal provided. So if anyone could come up with a way of providing people with the energy but not the coal, they'd be on to a winner. The answer was not a new source of energy, but two new vectors: gas and electricity.

Across the towns and cities of Victorian Britain the gas-ometer – those vast cylindrical repositories of 'town gas' – became a familiar totem of industrialisation as the gas-coke cycle became a chemical industry in its own right. Within each gasworks coal was heated in a vacuum so that the vola-tile alkanes could be boiled off and separated into gas and coal tar; the remaining solid, coke, was used in metal pro-cessing. A piped gas infrastructure was developed to provide street and factory lighting, allowing workers to get to the shop floor during the dark winter months and keep them operating profitably once they were there; through the night if necessary. Street lighting is something we take entirely for granted today, so it's worth remembering that these were the days before gasoline and the internal combustion engine, when the swiftest form of transport was horse-drawn. Gas was also used to provide domestic light in more affluent homes and districts. The vestiges of this Victorian system are still with us today. Our modern gas pipes follow the same paths around town, using the same principles of distribution, and while the 'natural gas' we use is different – essentially pure, odourless methane imported from Norway and Russia, rather than a locally produced mix of methane and other calorific gases such as hydrogen and carbon monoxide – it has sulphur compounds added so that it smells just like the old town gas.

In America, the creation of an electricity supply was driven by Thomas Edison, the inventor of the incandescent light-bulb. Edison was like a 19th-century Steve Jobs. Regarded today as a brilliant inventor, he was in fact a brilliant adapter of other people's inventions, with a shrewd business-eye for the big chance. Edison had grand plans to electrify the whole of New York using his own electricity-generating system called direct current (DC), but he wasn't the only person trying to build an electricity business. His bitter rival in this endeavour was initially the equally ruthless J.D. Rockefeller. Edison's lightbulb was only the latest in a series of innova-tions involving electricity. Heaters had already been invented

and were providing power and warmth to homes and factories. Oil magnate Rockefeller was attracted to this new means
of distributing power, largely due to concerns about losing
some of his energy market share.

Edison hired talented European engineer Nikola Tesla
to develop a DC system that could supply the new electricity to downtown New York. Edison's plan for a network of
wires on poles high above the city, which could carry power
to factories, homes and offices, was proving impractical. The
problem with DC is that most of the energy generated at
source is lost (as resistive heat) while it's being transported
through the wires to its destination, which makes it impractical for them to stretch more than a few hundred metres. Ever
the pragmatist, Edison's solution was simply to build a generator every couple of blocks and hang the expense. Tesla,
however, became enthralled by the possibilities of a different
system for supplying power: alternating current (AC). AC, as
the name implies, works by constantly reversing the direction
of a current flowing in the circuit back and forth. In DC, the
current travels in only one direction. The big advantage of
AC is that, because the electrons in the wire are not moving
very far in either direction, the current is very low in comparison to DC at the same power, so very little is wasted as heat.
The upshot is that AC can therefore be transported at high
voltages over very large distances.[2] At the end of its journey
AC can be 'stepped up' or 'stepped down' as needed, giving
it a high current only at the point where it's required rather
than across the entire length of the cable.

With all the advantages of AC, one would imagine Edison
would have been equally attracted to it. However, lacking
the mathematical background of Tesla, he found the science
behind AC impossible to understand properly. Nor did he
particularly feel inclined to do so, since he had expensively

---

[2] Power = voltage × current, but resistance (and therefore loss of
power) depends only on the current. This means that high power can
be generated with low currents and very high voltages.

geared his entire Edison Electric Company to a DC system. This at least provided him with some short-term advantages. At the time there was no AC motor available and Edison had already invented a meter that allowed his customers to be billed in proportion to their consumption. Unfortunately for Edison, though, his reticence towards AC quickly began to manifest itself as antipathy towards his star employee. The pair had a terminal falling out when Tesla discovered that some of his inventions had been patented in Edison's name.

Tesla left, taking his ideas with him, and the so-called 'War of Currents' began. Edison focused entirely on the singular advantage of DC over AC – that it is, in theory, a safer system. After initially partnering with Rockefeller, Tesla and his subsequent backer George Westinghouse won out. As a result, the AC power distribution system we use today is based on the principles developed by Tesla over 100 years ago. The biggest advantage of AC is that one massive power station can send its power out at high voltage (up to 750 kilovolts) across vast distances. This voltage can then be stepped down for use industrially and domestically, by transformers in substations. Today, almost all electricity is supplied in this way: typically at 110 V and 60 Hz in North America and Asia and 240 V and 50 Hz in Europe.

Rockefeller needn't have worried: electricity and town gas spelled the end for petroleum as far as lighting and heating was concerned. But oil had far too many advantages to ignore. It too has a very high energy density but because it's a liquid, it's particularly easy to store and move around. Unlike coal, oil can be pumped, piped or poured, which meant it was the perfect fuel for the newly invented internal combustion engine. A third and final energy infrastructure was created to supply liquid fuel to industry and consumers.

It's easy to see why we fell in love with petroleum, but the unprecedented wealth enjoyed by Rockefeller and his fellow oil barons is testament to the fact that this is one love that undoubtedly became an addiction. Cheap, liquid fuel made it cost-effective to transport manufactured items from their place of origin to markets all over the world. What began as occasional luxuries for the few have become every-day staples for us all. Our homes today are filled with goods made in South-east Asia; we eat African fruit and chocolate, drink Australian wine and South American coffee and drive German, Japanese or Korean cars. It's a sobering thought that 27 per cent of the world's energy budget is expended on transporting these goods across the planet. The vast majority of this activity is fuelled by petroleum.

Today we still rely on the same three vectors: oil, gas and electricity (which can be produced in any number of ways, but usually by burning fossil fuel) to meet all our energy needs. The parallel and complementary supply systems that deliver them are superbly optimised: energy is always there whenever we need it wherever we need it. Yet despite this ubiquity, fuel has almost completely disappeared from our lives. Gas and electricity are piped and wired directly into our homes; where the gas comes from, or how the electric-ity is generated, is a mystery that few of us are interested in uncovering. Even filling the car with petrol – the closest we come to actually handling the fuel – is a necessarily sterile process; we could be putting anything into our tanks (as any-one who's mistakenly put diesel into a petrol engine or vice versa will be able to attest) and the origin of the fuel itself is unknowable. There are obvious safety benefits in not hav-ing stockpiles of fuel scattered here and there – certainly our towns and cities are generally much cleaner environments than they were even 50 years ago – but in the process we have become disengaged from the realities of energy gener-ation. Out of sight is out of mind and our profligacy when it comes to consuming energy is simply because we take it for granted. Our Elizabethan forebears understood the effort required to maintain a supply of firewood through the winter months and would undoubtedly have been less inclined to pen letters of complaint about blots on the landscape created by wind-farms and power stations.

Escalating fuel bills are just the beginning of our problems. Before long gas, oil and coal will become obsolete, and the big question is: can we maintain the vector infrastructures with the alternative energy? In other words, how can we keep the wagons rolling and home lights and fires burning? But before we can answer that there's one more issue to consider, because the challenges presented by life after fossil fuel don't stop with power for homes, industry and transportation – fossil fuel drives our entire system of agriculture too. As a result, there's also the small matter of how we are going produce enough food to fill 9 billion bellies.

7   Food, Glorious Food

'Abundance does not spread; famine does.'
Zulu proverb

'The Shepherdess', an 1889 painting by William-Adolphe Bouguereau, is the epitome of the pastoral idyll. A beautiful young peasant girl on the verge of womanhood appears to be returning from a day in the fields and is posed balancing a stick across her shoulders. The girl is barefoot, symbolising her fiscal poverty, but in contrast her figure is full; she is comely, well-fed and in good health. Her work now complete, she looks tired yet unsullied by the rigours of a day's labour in the field. Her expression is one of indifference, as if she has more important matters to attend to than having her portrait painted. In the background we see sleek oxen grazing in the field. The countryside itself, like the girl, appears to be scrupulously clean. The painter's message is clear: here is someone enjoying a much simpler but far nobler and therefore better life than you. Our attraction to the girl is a metaphor for our desire to escape the complexities of urban living; of a deeply-held yearning for a return to the simpler, pastoral way of life.

The pastoral movement of the 19th century is responsible for the enduring myth that, prior to the industrial revolution, the lives of ordinary people living in rural communities were simpler, better and happier. The notion of a noble peasant community, completely in tune with nature and eking out a sustainable existence based on organic farming, is a fallacy. Certainly, the objects of the pastoralists' fantasy – the late medieval peasantry – would not have shared their view. These people endured rather than enjoyed their lifestyle and would be quick to disabuse the pastoralists of their misconceptions. Farming was a hard and dirty business and the only reason they lived like that was because there was no alternative.

❋

Britain's period of slash-and-burn deforestation was over long before the energy crisis of the Elizabethan period. By the time of the Norman invasion, the native Anglo-Saxons were already exploiting as much arable land as possible, with around a third of the nation's 39 million acres (nearly 16 million hectares) used for cultivation (around the same as today) and a further third for pasture. The remainder consisted of wilderness, forests and human settlements. Throughout the Middle Ages, the vast majority of Britain's inhabitants – thanks to the Black Death, they fluctuated between 2.7 million and 4 million – were engaged in agriculture. Most were subsistence farmers tending narrow strips of land, held in common, with all produce split between the farmer and land-owner. These farmers were tenants, not landowners (usually title was held by either the church or some member of the aristocracy), but they were effectively tied to the land, handing over a high proportion of produce and labour (usually two to three days each week) in return for their tenancy.

The system of cultivation, refined over hundreds of years, worked by rotating crops. Typically one field would be used to grow a cereal, another would feature a staple vegetable like beans or peas, while the third would be kept fallow or used as pasture to feed animals. Nearly everybody kept some live-stock to provide additional food, but also as a source of goods or materials that could be traded to provide extra income for items that people didn't, or couldn't, make for themselves. Animals also played an essential role in regenerating the land, grazing freely on pasture during the day and spreading their manure over fallow fields during the night-time.

Overall, the average peasant diet of the time rates well against modern nutrition standards: low in fat, high in fibre. The daily staples were made up of cereals – wheat, oats, barley and rye – which could be boiled into soups and stews, baked into bread or brewed into beer. In such frugal times, the amount of ale consumed might surprise us – up to a gallon a day was not unusual – as might the fact that everybody drank it, even the children. But the alcohol content of the common 'small beer' was very low, and with no knowledge of germs or sewage treatment, placing your trust in the local water supply was tantamount to a death sentence. Medieval diets lacked vitamins A, C and D and were not high in calories, making the regular beer-drinking something of a necessity. Yet despite this foundation of healthy eating, the seasonal fluctuations in the availability of food were enormous and very long periods of poor nutrition, not to say borderline starvation, were the norm.

The medieval world was usually a very hungry one. Winter was the grimmest time of all, and stockpiling enough food to feed your family and your animals during the non-growing season must have been a profoundly stressful experience. Whenever possible, livestock herds were cut down through sale or slaughter. Without refrigeration, meat would be salt-cured and stored in barrels in an attempt to preserve it. Not that there was any meat for most people. There was no fresh dairy or cereals either. Cattle that were not culled were fed a subsistence diet that caused milk production to dry up, and farmers rarely kept grain reserves above levels required for the following year's seeding. Sustenance during this miserable time was provided by pottage, a thin gruel of whatever vegetables were around – onions, leeks, colewort, lentils – flavoured, if you were lucky, with a handful of herbs. In all, it was an extremely precarious existence. A bad storm, cold snap or dry spell could seriously reduce the crops available for winter, with little recourse to an economic solution during these frequent shortages.

Endemic malnutrition caused the people of medieval Europe to decrease in body size. Stunting is an evolutionary response to food shortages designed to bring body size into line with the calories available. It can occur if the mother is malnourished during pregnancy, or the individual before the end of adolescence. Once stunting has occurred, improved nutritional intake later in life cannot reverse the damage. Average heights for men fell from 173cm in the 11th century to 165cm by the middle of the 17th (in comparison with 175cm today). However, limiting body size in this way reduces not only the calorific intake required, but also life expectancy and cognitive development. The medieval population were not only smaller than we are, but on average they were less intelligent and shorter-lived too. From time to time regular shortages became more profound and famine occurred. Historically, people living on every continent have experienced prolonged periods of chronic food shortages. These famines happen whenever demand for food outstrips agricultural yield and there is a lack of resources – weather, labour, money or transportation – to grow more. Famines can last for many years. In medieval times, without the infrastructure to transport food over distances, they provided a highly effective, if unwelcome, limit on the maximum sustainable population.

The British agrarian revolution, which began in the 15th century and continued until around 1900 (making it more of an evolution than a revolution), succeeded in breaking the prevalent cycle of food scarcity. The obvious way to get more food is to increase the area under cultivation. For a medieval farmer, this could be achieved simply by reducing the amount of land left fallow, which could be as much as 30 per cent at any one time. The problem with this strategy is that the fallow stage was essential, because it allows the soil to recover its fertility. Improvements in drainage and soil management, together with a better understanding of the crops themselves, led to the development of a much more effective four-field rotation system, with turnips and clover added into the mix to help restore vital nutrients, eliminating the need for soil to be left fallow for a year. To manage the land properly, animals and agriculture have to be separated. Domestic livestock are just as happy to eat beans, pulses and cereals as they are fresh pasture, so keeping the animals under control required the hiring of labourers in the form of herdsmen or shepherds. Initially animals were put into small enclosures in the evening, usually to ensure they spread their manure there, but the wider benefits of separating them soon became obvious to landowners: keeping them fenced in all the time meant they required less looking after. The process of permanently enclosing animals started to pick up after the Black Death, which had caused labour costs to rocket. As well as being more efficient, enclosure meant that animals could be further separated and selectively bred through several generations to develop particular desirable features – such as woolliness, meatiness or milkiness.

With coal now fuelling a burgeoning metalwork industry, agriculture became increasingly mechanised. Advanced iron ploughs and inventions such as the seed drill further reduced the amount of labour required to till the land. These developments meant that productivity could be maximised by a small number of paid labourers working on the land, rather than by a community of subsistence farmers living off it. Together with enclosure, these changes also made it impossible for subsistence animal husbandry to continue, as villagers lost not only their land but also their grazing rights. Those unable to secure work were left unemployed, with many migrating to the towns and cities in search of work in the emerging factories of the industrial revolution.

Throughout this period, agriculture, industry and scientific innovation developed in lock-step. Without produce to feed the burgeoning population, now living in towns and cities rather than in farms and villages, neither the industrial nor the scientific revolutions could have proceeded. Without the technology provided by industry and the corporate finance structures that emerged to provide it with capital, agricultural yields would not have increased. Both in turn relied on the scientific breakthroughs and ideas of the Enlightenment – itself a product of the new, well-connected and well-fed urban intelligentsia – and the means to proliferate them. The printing press had made mass communication possible, and news of the latest technology spread rapidly through an increasingly literate Europe. In effect, each revolution supported and relied on the others in equal measure – and farming, technology and commerce have remained inextricably connected ever since.

The impact of the agrarian revolution on the British population during the 19th century foreshadows the global explosion we are living through today. In 1800 there were fewer than 9 million inhabitants living in Britain; by 1860 there was double that number, and when Queen Victoria died in 1901 her successor, Edward VII, found himself reigning over 32.5 million British subjects. This was almost a four-fold increase and it affected every area of the economy. Not only did these additional people require the production of four times as much food as at the start of the previous century, but more clothes, leather, carriages, horses, houses, furniture and all the other accoutrements of a modern life. Fortunately there was plenty of energy available to drive the necessary economic growth and meet this demand, while continually improving transportation and communications were reducing costs and leading to fully globalised logistics. Yet despite all these advances, agriculture simply could not keep pace with the demand. There was a ceiling on how much domestic farms could supply just by more efficient use of the land, mechanisation and complementary crop rotation. Across the industrialised world, populations were growing ten times faster than food production. The practical answer was again to increase the amount of acreage under cultivation and pasture. And if you can't use your own land, then use somebody else's.

With Britain's farming at its functional limit, the additional space required was provided by the colonies in North America, Africa and Asia. Nineteenth-century Britons were fed by the empire. Many British dominions (and also its former colony, the USA) had vast tracts of land with arable and pastoral potential. More than enough as it turned out.[1] Soon surplus produce from farms in India, Jamaica, Australia, Kenya and the rest of the empire was being sold to other industrialised nations, which over time came to be increasingly dependent on this imported food to make up shortfalls in domestic production. The pressure for new acreage continued unabated. By the early 20th century the world population was nudging 2 billion and agricultural shortages were once again a salient issue, only this time on a global scale. Without additional colonies to exploit, the only way to further increase yields was to improve the fertility of the soil itself.

The management of till fecundity was nothing new – it has been a preoccupation of farmers since the Neolithic – but an understanding of how and why fertilisers work was reached only in the mid-19th century. Plants grow by feeding on thirteen essential nutrients that are present in soil. When a plant dies and is left to decay, these nutrients are returned to the earth; but if the plant is removed (or harvested) before this can happen, then so are the nutrients. Land will recover if left fallow, but that can take years. So to speed the process up, fertiliser containing one or more of the missing essential nutrients can be added to the soil. The most important of the thirteen nutrients are the three macronutrients nitrogen (N), phosphorus (P) and potassium (K), which are all to be found in animal excrement. When animals – and humans – feed on vegetable matter, between 80 and 90 per cent of the nutrients originally present in the soil are excreted as urine or faeces. Spreading this manure on the fields is one way of returning these to the soil. By the late medieval period the

---

[1] In America, the newly opened Midwest had great soil – deep and fertile from thousands of years of being covered by self-mulching crops. Farmers used each field until it was exhausted and then simply moved on to an adjacent field. This systematic exhaustion was manifested as a loss of soil properties. As there was less organic matter to bind it together, the soil wouldn't hold water. It was this degradation that led to the famous dustbowl conditions in the 1930s.

benefit of spreading manure was widely understood (if not how it worked), and the process of harvesting excrement for use as fertiliser was well developed. Animal manure was a dispersed resource and its recovery often posed a logistical headache, with deposits lying here and there across hillsides, pastures and meadows.

Although animals cannot be easily persuaded to crap in the same place, let alone the right place, human beings can. Human manure (commonly known by the less objectionable name of 'nightsoil') contains decent levels of N:P:K (in a ratio of 0.51:0.1:0.23) and was used widely in market gardens and farms around towns and cities between the 15th and 19th centuries. The business of collection was the unfortunate responsibility of the 'gong farmers' – a job, one imagines, with few perks. Gong farmers were employed to remove human excrement from domestic privies and communal cesspits. Elizabethan towns were notoriously dirty, smelly places, but even their grubbiest citizens baulked at the sight of cartloads of raw sewage making their way through the market, so gong farmers were only allowed to ply their unpleasant trade during the night. In an early attempt at rebranding, those employed in this least attractive of occupations later became known as 'nightsoilmen', or simply 'nightmen'.

These 'organic fertilisers' have several major disadvantages aside from how they are collected. Firstly, they contain relatively low concentrations of the required nutrients and need processing to make them more effective, but if they are handled too much they can lose all their potency. Secondly, manure, whether human or animal, can only be used locally. If it's transported too far, the low nutrient concentration makes it uneconomical. Twenty tonnes of cow dung is required to provide just 100kg of nitrogen fertiliser – enough for one acre (0.4 ha). But if the manure is stored or transported over any distance beyond a few dozen miles, up to 40 per cent of the nutrient content will be lost before it's applied. In reality this means that an average-sized farm of 150 acres (61 ha) would require almost 5,000 tonnes of non-locally-sourced manure to be delivered each year. Shipping that amount today would require a fleet of 113 HGV lorries. Finally, the majority of the nitrogen in manure is locked into organic compounds, which can sometimes take years to break down. Despite these drawbacks, manure and other organic substances were pretty much the only kind of fertiliser available until the 20th century.

A major breakthrough occurred in 1827 when Justus Freiherr von Lieberg identified nitrogen as an essential plant nutrient. Known as the 'Father of the Fertiliser Industry',[2] Lieberg popularised the Law of the Minimum, which states that plant growth is limited not by the total amount of nutrients available, but by the level of the scarcest nutrient. In practice the limiting nutrient is usually nitrogen, so Lieberg deduced correctly that adding it to the soil – in the form of ammonia – would have a profound effect on fertility. Unfortunately, knowing what to do is not the same as knowing how to do it, and the key question was: where is the ammonia going to come from?

In the early 1800s ground bones – usually sourced from slaughterhouses or battlefields – were being used widely alongside manure. After 1830, with nitrogen now recognised as the 'active ingredient', these bones were dissolved in sulphuric acid and mixed with a substance called nitrate of soda (sodium nitrate) to produce liquid ammonium nitrate, a much more effective fertiliser. Sodium nitrate is a naturally occurring salt with the formula $NaNO_3$. It's also known as Chile saltpetre, in part because of the vast deposits found in the Atacama desert on that country's coastline, but also to distinguish it from ordinary saltpetre (potassium nitrate). Prior to Lieberg's discovery there was very little demand for sodium nitrate. The first shipment to Europe arrived in England some time between 1820 and 1825, but there were no takers and the cargo was summarily dumped overboard

---

[2] A title we suspect, given its gong-farming-and-nightsoil heritage, few coveted.

to avoid excise duty. However, by 1859 nitrate was big business, with the UK alone importing 47,000 metric tonnes each year. The trade was ultimately monopolised by a flamboyant Yorkshireman called John Thomas North. Born in Leeds in 1842, North was an engineer sent to Chile in 1869 to deliver a consignment of mechanical equipment. While there he recognised the potential in the undeveloped saltpetre field in Chile and neighbouring Peru. This is by far the world's largest naturally occurring source of nitrate: a single area of prehistoric bird guano, approximately 220 miles (354 km) long and two miles wide, very near to the surface and therefore easily extracted. With remarkable ease, North set about buying up as much land as he could. Nitrate was initially carried to the sea by armies of pack mules, many of which died of thirst along the way. North literally couldn't mine enough of the stuff and invested heavily in every technology that would increase production, setting up, among other things, an aqua system to provide water to his miners and a private railway that could ship the nitrate to Pacific ports in the vast quantities required to meet demand. Nitrate had to be kept dry, but in the holds lifted and dropped by Pacific and Atlantic swells, it reacted with the moisture in the air to release a vapour that made the sailors lethargic and was known to kill woodlice, rats and even the more unfortunate ships' cats.

☀

During the first decade of the 20th century, the worldwide demand for nitrogen-based fertilisers exceeded the amount that could be supplied naturally. Even the 2 million tons of nitrate that Chile was shipping each year by 1912 was not enough. As such, it was serendipitous that almost 80 years after Lieberg's discovery, another German scientist called Fritz Haber finally managed to create ammonia artificially. By doing so he provided yet another important application for the ancient sunshine trapped in coal and oil. Haber's industrial synthesis of ammonia from nitrogen and hydrogen has proved of fundamental importance to the modern world: greater than the invention of the aeroplane, nuclear power, television and the internet combined. Quite simply, the expansion of the world's population from 1.6 billion people in 1900 to today's 7.1 billion would not have been possible without it. Yet the man himself is one of science's most equivocal figures.

Born in Breslau, Prussia in 1868 into a middle-class Jewish family, Haber was undoubtedly a genius. Erudite and driven, he enjoyed a brilliant academic career and was awarded a Nobel Prize in 1918. Yet he is also known as the 'Father of Chemical Warfare', a title that was well earned.[3] It seems ironic that a man who devoted most of his time to conceiving inventive ways to kill people should also be credited with the scientific innovation that has allowed the Earth's population to quadruple in less than a century, but ammonia is a major component not just of fertiliser, but of explosives too. Haber's ambivalent character is highlighted by a probably apocryphal tale concerning a meeting with Einstein. The story goes that after attending Haber's demonstration of his newly developed chlorine gas, Einstein was so disgusted by the application of science to mass murder that he rejected the offer to convert his citizenship from Swiss to German and refused to sign the 'Manifesto to the Civilised World' – a list of prominent German scientists, artists and academics who supported Germany's involvement in the First World War and its invasion of neutral Belgium. Whether true or not, the story illustrates a conflicted personality; and indeed, during the war, Haber acted as a consultant to the German War Office, organising gas attacks and devising defences against them.

But his achievements were little short of miraculous nonetheless, as chemists had been trying to synthesise ammonia for almost 100 years. Haber's breakthrough was the direct

---

[3] Another nickname, but one that contrives to make even Lieberg's alias sound almost appealing.

reaction of nitrogen with hydrogen to make the substance using a high-pressure metal catalyst within an innovative double-walled steel converter of his own design, to contain the pressurised gases. He gave the first public demonstration of his process in the summer of 1909. The German chemical company BASF immediately recognised his patents' value and bought them off him, and also agreed to pay a royalty on any ammonia they produced. There was still some way to go before this lab-bench curiosity could become production plant. The company charged one of its star employees, Carl Bosch, with the onerous task. Bosch's contribution to what became known as the Haber–Bosch process – enough to also win him a Nobel Prize – was to design and construct an industrial-scale system capable of withstanding the high pressures and temperatures involved. There were several practical problems to be overcome. Haber's prototype had used catalysts made of osmium and uranium, both of which were prohibitively expensive barriers to industrialisation, and cheaper alternatives had to be found. There were also the significant matters of constructing a safe, high-pressure blast furnace and developing a cost-effective means of purifying the gases used to synthesise ammonia. Step by step, Bosch and his team overcame them all, and in 1916 ammonia was being produced in commercial quantities at BASF's plant in Oppau, Ludwigshafen.

Bosch's industrial version of the process strips out hydrogen from methane (CH_4, natural gas) via carbon monoxide, which is highly toxic, using a pair of reactions called steam-reformation and the water-gas-shift reaction. All of the carbon monoxide is recycled, but lots of energy is consumed in liberating the highly reactive hydrogen and converting the methane into carbon dioxide. Within the converter, ammonia is produced by reacting one nitrogen molecule with three hydrogen molecules to make two ammonia molecules at pressures of 200 atmospheres (3,000psi) and temperatures of 500°C.

The widespread availability of ammonia during the

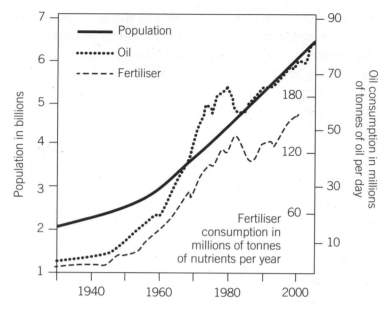

Fig. 3: World fertiliser consumption, oil consumption and population growth during the 20th century. (Sources: International Fertilizer Industry Association and UN Population Division)

inter-war years led to further innovations that dramatically affected soil fertility. Ammonium sulphate was synthesised in 1923, nitro-chalk in 1927 and ammonium phosphate in 1931, each one offering a more concentrated and economical fertiliser than before. By 1932, Haber–Bosch was providing more than 50 per cent of nitrogen in all fertiliser and it remains vitally important today. The ammonia it produces sustains two-thirds of the Earth's population, and more than half of the protein in each of our bodies contains nitrogen that was originally fixed by the process. Haber–Bosch is certainly an effective way of producing ammonia, but it's energy-expensive. It accounts for almost 3 per cent of the global annual supply of natural gas.

The use of ammonium nitrate fertiliser, whether organic or synthetic in origin, increased productivity by allowing both the amount of arable land under cultivation to be expanded and for that cultivation to become more intensive, but it had little impact on crop yields themselves. Throughout this period harvests rarely exceeded 2.0 t/ha (metric tonnes per hectare). Increasing yields required new strains and species of plant that could take advantage of the now abundant supplies of nitrogen and other nutrients by converting them into seeds and roots rather than stalks and leaves. This major breakthrough in yields took place during the 1950s, when plant breeding underwent a 'green revolution', leaving the farm and entering the laboratory. Throughout the 20th century the science of agriculture became more industrialised, incorporating not just developments in fertiliser, plant and machinery, but seed production, irrigation and bespoke pesticides and herbicides. However, without question the most important development of all was the emergence of a variety of high-yield plant cultivars after 1950.

The man credited with leading the research that resulted in the green revolution is Norman Borlaug. Born in 1914, Borlaug was a real son of the corn. He grew up in the Midwest grain belt, where he worked on the family farm near Cresco, Iowa. His grandfather persuaded him to leave the farm and gain an education, telling him: 'You're wiser to fill your head now if you want to fill your belly later.' He became a plant pathologist and microbiologist and in 1944 was approached by a new body dedicated to agricultural research, funded by the Rockefeller Foundation and sponsored by the Mexican government. The aim was, for the first time, to study genetics and plant breeding in the context of a host of external factors: pathology, entomology, agronomy, soil science and cereal technology. Borlaug accepted and moved his family to Mexico City, where he was installed as project leader at the Cooperative Wheat Research and Production Program (CWRPP). The goal of the project was simply to boost domestic wheat yield at a time when Mexico was importing most of its grain.

Unlike Haber, Borlaug's personal aims were not exclusively scientific. He also had the humanitarian aim of feeding the hungry people of the world by providing, as he said presciently at the time, 'a temporary success in man's war against hunger and deprivation'.[4] During the sixteen years he spent at the project his team enjoyed a series of spectacularly successful results with the high-yield, short-straw, disease-resistant varieties of wheat they developed. Their key break-through was the development of crop traits that would not have been favoured by natural selection. Borlaug worked out that low-stature wheat, and other crops that produce very high yields, were able to do so by allocating more resources to the grain at the expense of height and competitive abil-ity. He began to artificially select for those traits. In 1963 the focus for Borlaug's work became the International Maize and Wheat Improvement Centre (CIMMYT), which, from its base in Mexico, worked initially with the Indian govern-ment to enable the diffusion of this new technology into the field. Today, little more than 40 years after CIMMYT wheat hybrids were first used in the Punjab, yields of the crop have increased from 0.5 t/ha to more than 4.0 t/ha – a seven-fold jump.

Following the Mexican-led success with wheat and maize, the Ford and Rockefeller Foundations persuaded the Filipino government to establish the similarly motivated International Rice Research Institute (IRRI) and do the same for the planet's most-consumed staple. The IRRI breeding programme was just as fruitful as its forebear and in 1966 one of the strains they had developed became an entirely new cul-tivar, IR8. This new rice variety required the use of fertilisers and pesticides to thrive, but produced substantially higher yields than the traditional varieties. Annual rice harvest in the Philippines increased from 3.7 t/ha to 7.7 t/ha in just two decades. The switch to IR8 rice made the Philippines a rice exporter for the first time in the 20th century. India, the world's second-biggest producer, followed suit using a new IR8 hybrid that yielded 5.0 t/ha without fertiliser but almost 10.0 t/ha under optimal conditions. This was ten times

[4] Note his use of the word 'temporary'.

the yield of traditional cultivars and IR8 was immediately dubbed 'miracle rice'. Production has continued to increase ever since. During the 1960s average yields in Indian paddy-fields were never more 2.0 t/ha. By 2000 production was more than 6.0 t/ha. The increase in supply meant that prices fell. In the 1970s rice cost about $550 per tonne; by 2001, it cost less than $200 per tonne.

Like Bosch and Haber before him, Norman Borlaug was honoured by the Nobel Academy in 1970. A telling difference, though, is that Borlaug was awarded the prize for Peace. His acceptance speech particularly appeals to the chemist within us.

If the high-yielding dwarf wheat and rice varieties are the catalysts that have ignited the green revolution, then chemical fertiliser is the fuel that has powered its forward thrust.

The responsiveness of the high-yielding varieties has greatly increased fertiliser consumption. The new varieties not only respond to much heavier dosages of fertiliser than the old ones but are also much more efficient in its use.

The old, tall-strawed varieties would produce only ten kilos of additional grain for each kilo of nitrogen applied, while the new varieties can produce twenty to twenty-five kilos or more of additional grain per kilo of nitrogen applied.

By the time of his death in 2009, Borlaug's pioneering work to end world hunger had led to a trebling in cereal production across the developing nations. Yields of rice, maize and wheat have also increased through roughly equal contributions from developments in irrigation, fertiliser, and seed design. The green revolution of the 20th century had two profound effects on the Earth's human inhabitants. Not only has the population grown, but so have the people. The human body has changed more in the past 100 years than it did in the previous 50,000. Adults today are on average 50 per cent heavier and 4 inches (10cm) taller than they were in 1900, returning, height-wise at least, to the size of our hunter-gatherer ancestors for the first time in well over 10,000 years.

※

While it's undoubtedly good news that agricultural output has grown so dramatically, you never get something for nothing, and the downside to all this success is that the energy required to produce these crops has increased at an even faster rate. Today more net energy than ever before is needed to grow each kilo of food, because the new high-yield varieties rely not only on chemical fertilisers, which are themselves energy-intensive to produce, but also on pesticides and herbicides. So much so that modern agriculture is now entirely dependent on the petrochemical industry. The negative implications of so intimate a relationship are clear: if the oil runs out, then so will the food.

Equally apparent are the effects of the green revolution on global food security. Yes, the world population has grown by more than 4 billion as a direct result, but without high-yield varieties (HYVs) there would have been widespread famine and higher levels of malnutrition; the average person in the developing world now consumes roughly 25 per cent more calories per day than they did prior to the green revolution. Again, this achievement comes heavily caveated: agriculture now operates in a way that puts an enormous strain on the wider ecosystem by using crops that are wholly dependent on our intervention for their survival. This lack of security is the green revolution's unwelcome legacy, and is evidenced by the main criticism of modern agriculture – that a global benefit has been achieved through a series of local deprivations. Subsistence-oriented farming has been abandoned in many areas, with arable land geared instead towards the production of grain for export or animal feed. For example, in India much of the land formerly used to grow pulses that could feed the local population has been given over to wheat production. However, wheat doesn't make up a significant portion of the native diet, so much of this new harvest is exported. Furthermore, in Africa the impact of the green revolution on the diet of the indigenous people has been as good as non-existent. In contrast with the rest of the world, food intake per capita across the continent is down 20 per cent on 1960 and average crop yields are on a par with late-16th-century Europe at just 1 tonne per hectare.

Important it may be, but we rarely see or hear discussions about global food security in the media, perhaps because the issues appear removed from our day-to-day experience of full fridges and supermarket shelves groaning with produce. The concerns we hear about modern agriculture are invariably environmental: its reliance on chemicals and the loss of bio-diversity involved in putting more land to arable and pastoral use. Certainly the widespread application of herbicides and pesticides, when used inappropriately or indiscriminately, can and does have profound effects on both poorly-protected workers and the local fauna and flora. However, the anxieties expressed most often are about the impact of these chemicals on the health of consumers. (And in this context the word 'chemicals' is always used pejoratively.)

Large-scale monoculture (single species) with heavy fer-tiliser application and no crop rotation does undoubtedly lead to loss of biodiversity both locally and globally. Local biodiversity is often lost because traditional agricultural sys-tems had practices that preserved the wild fauna and flora, such as hedgerows full of small birds and rodents, which in turn support the larger birds and mammals that prey on them. The pressure to create yet more farmland has led to the destruction of millions of hectares of rainforest to pro-vide both cereals and protein (from either soya or meat), but more insidiously, the new HYVs, tolerant of soil conditions that previously prevented agriculture from happening at all, have led to the loss of steppe and savannah, which is now economically and practically capable of being converted to farmland as a consequence. This form of habitat loss is the primary cause of the degradation of global biodiversity.

To some extent, the organisations responsible for these problems have reacted to mitigate the effects of unsustainable farming practices. The Consultative Group on International Agricultural Research (CGIAR) plays a major role in collecting, characterising and conserving plant genetic resources and maintains over 650,000 samples of crop, forage and agroforestry genetic resources in a variety of seed and gene banks around the world. As well as running projects dedicated to the production of all the major food crops, special attention is paid to Africa in an effort to understand why initiatives that have been successful elsewhere don't necessarily work there. As well as traditional, indigenous African crops like cassava, the New Rices for Africa project (NERICA) combines the high yields of Asian strains with African strains' resistance to local pests and diseases.

To all the critics of increased food production Borlaug had this to say:

> Some of the environmental lobbyists of the Western nations are the salt of the earth, but many of them are elitists. They've never experienced the physical sensation of hunger. They do their lobbying from comfortable office suites in Washington or Brussels … If they lived just one month amid the misery of the developing world, as I have for 50 years, they'd be crying out for tractors and fertiliser and irrigation canals and be outraged that fashionable elitists back home were trying to deny them these things.

And we cannot help but agree.

Despite all the criticism, what remains a fundamental truth is that the 20th-century world could not solve its four-fold increase in population in the same manner as 19th-century Britain: there simply wasn't another planet available to colonise and exploit. The Haber–Bosch fixation of nitrogen, together with Borlaug's green revolution, provided enough food using what land is available. In China's Jiangsu and Hunan provinces, rice–wheat rotations, together with double-cropping, can produce protein at the rate of 800 kg/ha, whereas European wheat monoculture can produce 700 kg/ha. This is between two-and-a-half and four times more than the output of the most productive traditional agricultures, which rely on the extensive recycling of organic matter and the planting of legumes. Modern fertilisation methods can feed 35 to 45 people per hectare compared to just 12 to 15 using the best organic system. In theory, the most intensively farmed land on Earth can feed 50 people per hectare. In practice, those people would have to be Iowans getting 900 kg/ha of protein by rotating corn and soya beans. They would need to subsist on a strict vegan diet based exclusively on these two foodstuffs. An impractical solution at a local, let alone a global level.

To feed the world's growing population will require a 50 per cent increase in food production by 2030. This will have to be achieved with less water, less land and fewer inputs of fertiliser, herbicide and pesticide. Agriculture will have to become sustainable and much less energy-intensive. It will have to cope with the uncertain consequences of global warming and other environmental changes. Part of the solution will come from the development of further crops that are more resistant to pests and diseases, and cultivation systems that use sunlight, water and fertiliser more efficiently. Improvements in crop yield of the magnitude required to achieve sustainable, global food security will also require an increase in the efficiency of photosynthesis. One way to achieve this is to improve the intrinsic efficiency of the biochemical pathway of photosynthesis within crop plants, by introducing new traits found in other more efficient organisms – this means genetic modifications and consequently more GMs (genetically modified organisms). Another way is to readjust existing features of photosynthesis to optimise their productive capacity in a wide range of different environmental and climatic conditions – this also means genetic modifications and more GMs.

Food security depends on the sustainable and efficient provision of nutrients and water by soil and soil structure, all of which are enhanced by organic matter. Specific groups of microbes, notably mycorrhizal fungi, play a central role in controlling the flux of nutrients from the soil into plants and regulating the defence capabilities of crops. In natural ecosystems, mycorrhizas (microscopic fungi) supply the plant partner with mineral nutrients, acting as an extension of their root networks; this nutrient subsidy is paid for with carbon fixed through photosynthesis by the plant. Understanding the symbiotic nature of these plant/fungal interactions will allow us to create new varieties of crop plants that have a much lower impact on the environment. This will help us reconnect agriculture with its non-human ecosystem, making it less dependent on fossil-fuel-based inputs. There may well be additional opportunities for 'reversing' naturally selected adaptations and thereby breeding more cooperative plants, which would boost crop yields and help to alleviate global deficiencies in agricultural production. Here traits that favour individual fitness under natural conditions – such as aggressive root growth or horizontally positioned leaves – can be jettisoned in favour of those that improve overall yield by enhancing the performance of the entire crop population.

In combination, these developments will secure our future by changing the way we live and what we eat. And this change is crucial because the number of people we can support in the world is entirely dependent on how much they consume.

8   When the Explosion Stops

> 'Unlike plagues of the dark ages or contemporary diseases we do not understand, the modern plague of over-population is soluble by means we have discovered and with resources we possess. What is lacking is not sufficient knowledge of the solution but universal consciousness of the gravity of the problem and education of the billions who are its victim.'
> *Martin Luther King Jr, civil rights activist*

Throughout this book we have talked about a goal of sustaining a world population of 9 billion people, which will be reached around the middle of the century. The pertinent question at this point is: how sure can we be that the population will peak at 9 billion? Isn't there a good case to be made that it will simply continue to grow (in any one of a variety of consumption patterns) causing yet more deforestation, over-fishing, over-farming and untold environmental damage until there's simply nothing left?

Parking the doomsday scenario for a moment, it's certainly true that projecting future population growth is not an exact science. Indeed, even when it comes to estimating the current population there is no consensus. The United Nations Population Fund estimated that the global population reached 7 billion on 31 October 2011, while the United States Census Bureau, an equally august body, estimated that this milestone wasn't reached until 12 March 2012. Likewise, projections about the future high point vary accordingly, but these estimates should not be written off as wild guesses or stabs in the dark. There is a lot we do know, and the variance in the outcomes is largely due to what actions we might take in the future.

Notwithstanding the historical fluctuations (such as the Black
Death and great famine of 14th-century Europe), the global
population of humans has experienced continuous growth
since the first exodus from Africa. There is good reason to
believe that this growth has been hyperbolic – that is, the
time taken for the population to double halves with every
subsequent doubling. In practice, the initial doublings prior
to the Neolithic took tens of thousands of years, reaching a
minimum doubling time of just under 40 years during the
final decades of the 20th century (see Figure 4). The biggest annual increases – above 1.8 per cent – were experienced briefly during the 1950s and throughout most of the
1960s and 1970s. It should be noted that although the rate
of population growth has now slowed down, it's *still* growing. Total annual births were highest in the late 1980s at
about 138 million – simply because the babies born during
the period of peak growth were now old enough to start having children of their own. The birth rate is now expected to
remain constant at its 2011 level of 134 million, while deaths,
currently at 56 million per year, are expected to increase to
80 million per year by 2040, due in part to an increase in the
average age of the population during this period.

The upshot is that the time it takes to add a billion is
already stretching out, but given the huge number of people,
the population needs to start levelling off very soon. Current
projections all show a continued increase in population, but
also a steady decline in the population growth rate. The dis-
agreement is over the figure we end up with. It's vital to
remember that these are projections, not predictions. Were
we to take no action at all, the UN estimate does indeed
appear to concur with the doomsday scenario of continued
exponential growth reaching a wholly unsustainable 14 billion by 2100. At the other extreme, the most conservative
UN estimate suggests that we won't see much more growth
at all, peaking at 7.5 billion around 2040 before dropping to
just 5.5 billion by the end of the century. However, this is
not the good news it might appear, as in this projection the

Growth (billions)	Date	Doubling time (years)	Time to add 1 billion people (years)
0.25 to 0.5	1500	900	–
0.5 to 1.0	1804	304	100,000
1.0 to 2.0	1927	123	123
1.5 to 3.0	1960	79	33
2.0 to 4.0	1974	47	14
2.5 to 5.0	1987	38	13
3.0 to 6.0	1999	39	12
3.5 to 7.0	2011	44	12
4.0 to 8.0	2025	51	14
4.5 to 9.0	~2050	>50	25
5.0 to 10.0	Probably never!		

Fig. 4: Population over the past half-century showing an exponential rise until the late 1970s and then a gradual slowing down. (Source: UNFPA)

decline is triggered by a series of disasters linked to competition for scarce resources; in effect a Malthusian catastrophe of mass starvation caused by over population. That the population will reach 9 billion and then plateau is actually the UN's mid-range estimate. We don't know for sure whether this figure is correct, but what we do know is that the highest fertility rates are seen in the least developed countries, while the lowest fertility rates are found in the most developed nations. We are also seeing fertility rates dropping rapidly in the developing economies of Asia and South America. Global fertility is falling almost everywhere, but the reproductive momentum of such a large number, together with the proportion of young people in the world – 3 billion under the age of 40 – means that we are continuing to add in the order of 80 million souls per year.

How long we go on adding people, and at what rate, determines the number we will end up with, but there is also the question of where they will end up – these arriving citizens will not be evenly distributed. At the moment, six of the seven continents are widely inhabited on a permanent basis. The overwhelming majority of people, approximately 4.1 billion, live in Asia, the largest land mass containing the world's two most populous countries: India and China. Next is Africa, the poorest continent and also the one experiencing the most explosive population growth, with around 1 billion people in total. Europe and South America contain 733 million and 600 million respectively, while North America, the most affluent and also the biggest net consumer of natural resources, accounts for just 5 per cent of the total with around 352 million. The least populated region of all is Australasia with just 35 million inhabitants.

Implicit in all this is that raising people out of extreme poverty through economic development is the key to further reducing fertility rates. After all, the wealthiest nations in Europe and North America are experiencing a decline in their indigenous population, in the developing BRICs it's levelling off, but the 1.3 billion people living in the poorest countries in the world are neither benefiting from, nor indeed experiencing, this decline. Increasing the per capita consumption of this bottom group would improve standards of nutrition, healthcare and child mortality, which should provide the confidence to reduce family sizes. This certainly feels like it should be true, not least because we have our own experience to draw upon. In Britain during the years of post-war austerity, the average family contained 3.65 children. This had fallen to just 1.8 in 2011, some way below the replacement level of 2.1. Inevitably it's not quite as straightforward as it might appear: the actual number of people is not the whole story.

First of all, the level of consumption is just as important as population. Consumption is a measure of how peoples' wants and needs are assuaged. In economics, a need is something we have to have – like food, shelter or water – while a want is something we would like to have – like foie gras, a villa in the Algarve or a bottle of Perrier. Consumption comes in two forms: goods (or material resources) and services. Simply increasing levels of consumption across the board will have profound consequences for us all. Regardless of how many people there are, the Earth's capacity to meet the needs of its inhabitants is fixed: apart from sunshine there is only a finite amount of stuff. In addition, the exploitation of these limited resources has an environmental impact on the planet itself: loss of biodiversity and habitats, and climate change. The consumption of goods uses up natural resources, affecting supply levels and the environment (through extraction and processing).

Consumption in itself is not something we can live without, but it can be sustained over the long term only if renewable resources (such as wood, food or water) are replenished and non-renewable resources (metals, minerals and fossil fuels) are recycled, which is clearly not always possible. Consumption is correlated with population size and the level of economic development, which is why wide global variations in consumption patterns occur. More developed countries have the opportunity to consume more, while the less developed have their consumption restricted. In many parts of the world today, the basic needs for survival – in terms of water, food, energy and minerals – are simply not being met, but global consumption of water, food, energy and minerals is continuing to grow rapidly and disproportionately. For those living in poverty, the aspiration is merely for an adequate standard of living; in the developed world – where needs are generally assuaged – the drivers are more complex and tend to be based on ephemeral wants instead: possession of the latest mobile phone, a new car or a second home, for example. Adding people will inevitably increase levels of consumption, and so politicians shouldn't treat population and consumption as separate issues but rather as different sides of the same coin.

☼

We should acknowledge that our view that we are constrained

by environmental limits is not shared by everyone. Some economists argue that absolute scarcity does not and cannot exist, because market forces will push people towards alternatives or stimulate innovation to provide replacements; the Stone Age did not come to an end because there was a lack of stones. In these models, the seemingly parlous environmental state we find ourselves in is merely a transition before we develop the technological fixes to address the limits in natural resources. There is certainly some truth in this argument. For example, we have seen the effects that the market for oil has had on petroleum production. With the price consistently above $100 per barrel, it has become economically viable to start extracting oil from less accessible reserves – tar sands or deep-sea wells, for example. However, almost all environmental analysis suggests that not only are there genuine limits to the Earth's resources, but that the impact of scarcity is already being felt intensely in many parts of the world. Perhaps it's worth reminding ourselves of the famous words of Kenneth E. Boulding, who remarked: 'Anyone who believes exponential growth can go on forever in a finite world is either a madman or an economist.'

The extent to which people can substitute Commodity X with Commodity Y forms the centrepiece of debates over the extent of these ecological limits. In summary: can we develop technological solutions to scarcity or are the substitution possibilities limited in much the same way as the resources? We replaced cotton and wool with nylon and polyester, and horses with cars, so could we replace fossil fuels with solar energy, or meat protein with vegetable protein? Notwithstanding the technical and scientific knowhow required to achieve this, the answer relies entirely on whether these substitutions are created using renewable or non-renewable sources. Replacing the Gulf or North Sea oil with extractions from Brazilian or Canadian tar sands is a palliative rather than a curative measure. In this way the global economy can grow by exploiting its 'natural capital' for an extended period of time, but not indefinitely. To give another example, a small country could export its abundant supplies of natural gas and get rich in the process, but eventually the opportunity for these kinds of substitutions will disappear. However, there are grounds to be positive. Enormous additions to the sources of energy have been realised throughout history as we have moved from human and animal power, through wind, water, timber, coal, oil and natural gas to nuclear, solar and wave.

Secondly, the transition from high to low fertility rates is not driven exclusively by economic factors. In many parts of the world there are significant social, political and cultural barriers to overcome, and these influence the shape of society as well as the numbers. A growing population means there are more young people and a smaller proportion of elderly. In the richest countries that make up 42 per cent of the world's population, the average family contains less than 2 children; among the bottom 20 per cent, the number is between 5 and 7. As mortality rates and fertility rates decline, the proportion of elderly citizens increases. This is a process called demographic transition, and the pace at which countries move through it has a profound effect on birth rates and life expectancy. Economic and social development helps enormously, but so do improvements in education – particularly of women – and progressive family planning, especially where there is unmet demand for it.

Some economically developed countries such as South Korea and Iran, with enlightened family planning policies, have moved through the transition from high to low birth rates much more quickly than the US and Europe. Meanwhile in Cuba, Costa Rica and the Kerala region of India, high life expectancy has been achieved in spite of low incomes. Despite these positive examples, there is plenty of evidence to suggest that such a feat will not be repeated so easily elsewhere in the developing world. Despite improving life expectancy from under 40 in 1980 to almost 55 in 2011, Niger is experiencing one of the highest annual population growth rates in the world at 3.5 per cent. A quarter of all women under the age of 40 have ten or more children. In part, this is due to endemic poverty, low standards of health-care and education and unavailability of family planning, but the primary reason is the local culture of polygamous union and a widespread desire for large families: married men and women want an average of 12.6 and 8.8 children respectively. Even assuming the fertility rate falls to 3.9 children (to take the UN's median projection) the population of Niger will grow from 15.5 to 55.5 million by 2050, with a real chance of completely overwhelming food production and supplies of other essential resources. Averting this disaster will not be achieved by increasing domestic food production alone. International migration would be inevitable, which would simply export the problem to neighbouring countries, all of whom are struggling with similar, if slightly less extreme versions of the same issue.

Finally, population is not simply about the number of people. Factors such as age structure, rate of urbanisation, differences in culture, and migration are important considerations for determining policy. For example, population decline, in economic terms at least, presents an even greater challenge than growth, because its impact on age structure means that a smaller working population is available to support – and pay for – the increasing numbers who have retired. China is an extreme case: by 2040, for every Chinese worker there will be nine people who have reached retirement age. Meanwhile, for those living through it, ad hoc urbanisation in 21st-century Africa remains the same harrowing business that it was in 19th-century Europe. Africa is the least urbanised continent but contains the fastest-growing cities. Slow economic growth and poor urban planning – due to lack of finance and expertise – have resulted in a shortfall of economic opportunities and expansive areas of slum dwellings. Sixty-two per cent of Africa's urban residents live in slums or shanty towns – over 200 million people. Migration is less of a black-and-white issue. It's by no means intrinsically negative, and can have many benefits – offsetting skill gaps and shortages in the host territory and providing remittance to the country of origin. It really depends on the factors driving migration: often it's the groups most vulnerable to economic, political and social deprivation who migrate.

It's a different story in the UK where, despite a falling birth rate, the population is projected to grow from the current level of 62 million to 74.5 million by 2050. The growth is due largely to net inward migration. This migration will help to mitigate many of the negative economic consequences resulting from an ageing workforce and will also benefit the education system (as domestic demand drops off), compensate for skill shortages, and improve social mobility and the competitiveness of the labour market. Without migration 20 per cent of the European workforce would disappear by the century's mid-point, which would have profound social as well as economic implications. Studies consistently show that migrants make a net contribution to the British economy. Furthermore, around 50 per cent return to their homeland or emigrate elsewhere within five years, which means they have a much lower impact on the old-age dependency ratio than the indigenous population.

We can identify the key challenges that have to be addressed if we are to have a positive outcome when the explosion stops. First, it's essential that the 1.3 billion people living on less than $1.25 per day are raised out of poverty. Equally important is that consumption is reduced immediately across developed and emerging economies. Finally, population growth must be slowed and stabilised. The issue with this is that increasing consumption is fundamental to all of the current economic models that are based on producing growth: which, by implication, means using more, not fewer, natural resources.

Unfortunately – referring back to the point about the finite capacity of the Earth – there isn't really a choice. The complexity of our way of life, our socio-economic systems and the relative ineffectiveness of the traditional forces of innovation in isolation mean that the relationship between population, consumption and the environment is not straightforward, but the three are equally significant. The capacity of the Earth to meet our needs is fixed, but how long it takes us to reach those limits is a question of choice: about what is used and how it's used; about what is important and essential; about how we are going to live and how we expect everyone else to live.

The projections may vary considerably, but the fact remains that in the developed nations per capita consumption is already far above a level that can be sustained for everyone today, let alone tomorrow. There are 3 billion people living in the BRICs aspiring to the lifestyle enjoyed in the OECD, while 1.3 billion continue to live in extreme poverty. The reduction of consumption is the salient short-term priority. A sustainable, unequal world is both conceivable and achievable, but surely our goal should not become the creation of a future in which the extreme poverty of the majority becomes a necessity to maintain the lifestyles of a privileged few?

The Royal Society report *People and the Planet* concludes with a cogent plan of action. Slowing population growth in the least developed countries will alleviate many of the challenges, and family planning education and provision should be a fundamental health priority. Likewise, the reduction of consumption in the developed world – and particularly of non-renewable resources – is essential. This will require greater investment in renewable energy technologies and a commitment to waste reduction and greater efficiency, but also 'the systematic decoupling of economic activity from environmental impact.'

Achieving this without coercion will be a huge challenge. Wide-scale adoption of sustainable technologies would mitigate the environmental impact of consumption, and re-using equipment, recycling materials and reducing waste are essential, but a radical change in thinking and a commitment to the long term by our political leaders is also required. We will have to achieve a reduction in the material and environmental impact of economic activity, which must in turn become collaborative as well as competitive.

Accepting that we may be entering a period of post-growth is a frightening prospect – if we remain committed to traditional economic models and performance measurement. Capitalism is not a perfect system – it requires regulation – but was it really responsible for the mess we find ourselves in? Since 2008 we have witnessed a series of spectacular business collapses. The crisis within the banking sector was inevitable (ultimately because the banks' structure proved to be inherently flawed) but the institutions themselves were not really operating as capitalist bodies at all. Risk and return is a fundamental capitalist principle, yet it has been absent in the financial sector for many years. Yes, there were incentives to succeed, but there were no incentives not to fail. Risk-taking was rewarded regardless of the outcome. The prize for ruining a big company in the 21st century is a multi-million-pound pay-off. Senior executives announcing record losses look forward to gilt-edged severance agreements, golden goodbyes and enormous wealth.

Corporations in the public and private sectors are set up like dinosaurs – tiny brain, huge body – and they create a working environment where accountability and responsibility are in short supply. Rationalising the struggling financial institutions into a smaller number of state-owned mega-corporations is another palliative measure that will compound problems in the long term. Competition doesn't give you choice: it gives you winners. The downturn offers us the opportunity to learn from the mistakes of the past, and the lesson is clear: small, flexible organisations are the most transparent, efficient and ethical. We should be breaking up, not consolidating, but as Neil Sedaka said, that is very hard to do.

Living in Europe, the global population explosion has been very easy to ignore. Our countries don't feel particularly more crowded than they did 20 years ago (and indeed comparatively speaking they aren't), while, the occasional hosepipe ban notwithstanding, we don't experience shortages of food, water or energy. But that doesn't mean it's of no concern. In the 5th century the two empires of Persia and Rome, which had dominated the world from Britain to Afghanistan for centuries, were facing collapse. Their borders, heavily fortified to protect against marauding invaders, their large standing armies and complex socio-legal system were proving no match against the overwhelming number of economic migrants who simply wandered over in small groups in search of work, food and a better standard of living. Starving people don't hang around – they move about in search of food – just as the homeless look for shelter, the dispossessed for employment. It's up to us to decide whether the needy of the world are an opportunity or a threat, but either way it's everybody's problem to solve.

If population growth is the biggest social and economic issue facing humanity today, then the single most important technological challenge in terms of reaching a solution is energy.

9   What's So Good About Oil

'The currency of the world is not the dollar, it's
the joule.'
*Nathan S. Lewis, Professor of Chemistry, California
Institute of Technology*

It's easy to understand why we fell in love with oil. Naturally
occurring and energy-dense, crude oil (or petroleum) is a
mixture of hydrocarbons and other organic compounds
found in geological formations deep below the Earth's surface. Oil is liquid at ambient temperatures, which means it
has neither the storage nor transportation problems associated with natural gas. The long hydrocarbon molecules that
make up the bulk of crude oil can be converted into a variety
of shorter alkanes and alkenes, which are much more useful.
The process of refinement, called cracking, involves boiling
the oil to produce an array of consumer products with different properties: from petrol, diesel and kerosene to asphalt,
plastics and pharmaceuticals. Although oil has been used for
more than 4,000 years, it's only since the invention of the
internal combustion engine that we have really woken up
to its potential. Today oil accounts for 40 per cent of total
energy consumption in the UK and USA and fuels 90 per
cent of all domestic transportation. Yet oil is more than just
a fuel: it has a big part to play in agriculture as a chemical feedstock, while petrochemicals are the source of pesticides and herbicides. Oil is also the basis of nearly all plastics,
and our mobile phones, televisions and computers wouldn't
work without oil-based products such as liquid crystals and
insulators. Oil's unique properties make it, in almost every
way, the perfect fuel.

Almost every way.

Although petroleum is a naturally occurring substance
present in the environment in some parts of the world in

seeps and tar-pits, the process of extraction is a dirty business. Water and oil famously don't mix, but oil is actually extracted from the ground as an oil-in-water emulsion, which requires chemical separation using de-emulsifiers. This means that inevitably some oil is released into the environment, which often has profound effects on the local ecology. Most of our remaining reserves of oil are off-shore in deep-water sites, which makes extraction exponentially more difficult, accidents more likely to happen and restitution more difficult to achieve when they do.

The Deepwater Horizon macro oil spill, which occurred in the Gulf of Mexico during 2010, ravaged the marine fauna and flora across an 80-square-mile 'killzone'. It devastated coastlines and ruined the region's fishing industry. Around 62,000 barrels of oil per day were deposited over a three-month period before the leak was capped. Deepwater Horizon is the most recent example, but despite tighter regulation it's unlikely to be the last, as long as our thirst for oil remains unquenchable.

The burning of oil produces large quantities of carbon dioxide, which makes petroleum, together with coal, the biggest contributor to the increase in atmospheric CO_2 and the resultant global warming. The environmental cost of using oil (and coal) is enormous. Notwithstanding global oil and coal reserves, it's a price that we are unlikely to be able afford much longer, economically as well as environmentally.

*

What percentage of your income would you be prepared to spend to keep warm? The cost of heating homes and workplaces has risen significantly over the past decade. With 2011 seeing some of the largest annual hikes in living memory, there is little cause to expect this trend to be reversed any time soon. In the US one half of households spent 21 per cent of their total income paying energy bills in 2012, up from just 11 per cent of total income at the turn of the century. In the UK the main energy companies (the so-called 'Big Six') all

increased their prices for gas and electricity by an average of
17 and 10 per cent respectively. Low income is a key driver of
fuel poverty, and rising fuel bills hit the poorest members of
society hardest. In the UK a third of the poorest 20 per cent
of households are living in fuel poverty, which is defined as
a household that needs 'to spend in excess of ten per cent of
[its] income on all fuel use in order to maintain an adequate
level of warmth'. This is typically 21°C in the main living
area and 18°C elsewhere. Fuel is needed for other purposes
too, such as heating water, lighting, cooking and transport.
Escalating fuel costs are a symptom rather than the cause of
the problem. Increasing energy efficiency within the home
is vital, but unless we can find alternatives that make gas,
oil and coal obsolete, we will all be priced out of the market
long before supplies of fossil fuels are exhausted.

The big question is: can we maintain our current vector
infrastructures with the new forms of energy? In other words,
how can we keep the cars and trucks rolling, the lights and
fires burning, by delivering energy through pumps, pipes and
wires? To answer that, we first need to understand the factors
that have resulted in petroleum oil becoming the most successful fuel of all time. Oil gets a bad press, and much of it is
deserved, but we shouldn't ignore the fact that it has several
properties that combine to make it an almost uniquely useful
substance.

The Standard International (SI) unit of energy – the ability
to do work – is the joule, named after the man who established its definition. James Prescott Joule was born in Salford,
Lancashire in 1818. He was both a physicist and a brewer by
trade,[1] and his amateur status meant that his early attempts to
curry favour with the Royal Society met with haughty disdain.
His first papers were dismissed as the parochial ramblings of

[1] This relationship between beer and physics has been celebrated by
science students ever since.

a dabbling amateur. Joule was undeterred, and after noticing that burning a pound of coal produced five times as much energy as a pound of zinc, he turned his attention to the convertibility of energy. He speculated how much 'work' could be extracted from any one given source – and by endeavouring to find the answer, he devised what was to become his best-known experiment. In this, the work released through the action of a 'weight falling through a height' was used to turn a paddle-wheel in an insulated barrel of water, which in turn increased the temperature of the water. The friction and agitation of the paddle-wheel on the body of water caused heat to be generated. By measuring the temperature change of the water and the height of the weight's fall, Joule was able to determine the mechanical equivalent of heat. These findings appeared in his breakthrough paper, *On the mechanical equivalent of heat*, which was published in 1845. Joule had shown that motion and heat are mutually interchangeable: a given amount of work will generate the same amount of heat, provided the work is totally converted to heat energy; but also that, due to the Second Law of Thermodynamics, the same could not be said in reverse. Despite initial resistance, in part due to his dependence on suspiciously precise measurements, his ideas were finally accepted. Joule later went on to work with the esteemed Lord Kelvin to develop the absolute scale of temperature.

The joule, named in his honour, is equal to the energy expended (or the work done) in applying a force of one newton through a distance of one metre (N·m) or in passing a current of one ampere through a resistance of one ohm for one second. Our modern definitions of energy, heat, temperature and work all have their roots in Joule's experiment. The US/imperial units for energy and work include the horsepower hour (2.6845 megajoules or MJ), the foot-pound force (1.3558 J) and the British thermal unit (approximately 1,055 J). Electricity bills are calculated in kilowatt hours, each one equivalent to 3.6 MJ. Food is measured in calories, one of which is equal to the amount of thermal energy required to raise 1 gram of water by 1 degree Celsius (or 4.184 J). Even explosions are standardised, with 1 gram of TNT equal to 4,184 J. Although when we use them ourselves, the words 'power' and 'energy' are effectively interchangeable, to scientists and engineers they are completely different things. Power is the rate at which energy is converted from one form to another, and we measure this in joules per second or watts. In a power station, coal is burned to heat water, which can drive turbines to produce electricity. Here the potential energy of the coal is converted first into heat, then kinetic energy and ultimately into electricity: the amount that is generated per unit of time is the power. Power is measured in several different ways: other units include ergs per second, horsepower, metric horsepower and foot-pounds per minute. These are all simply different ways of expressing joules per second.

Energy density relates the amount of energy stored to the volume of the storage facility. For example, if we know the capacity of the fuel tank in your car, we also need to know what kind of fuel is in it before we can work out the vehicle's range and how many miles per gallon you are likely to get. The greater the energy density, the more energy may be stored or transported for the same volume. Your car may be able to run on either petrol, ethanol or liquid petroleum gas. The amount of fuel in the tank is the same each time, but the performance will vary depending on the fuel's energy density. Fuels with a lower density have less potential energy available to be converted into kinetic energy, but just because a substance has a high energy density, that doesn't mean it can be converted efficiently. Butter contains much more energy pound-for-pound than ethanol, but you'd struggle to get any sort of fuel performance out of a butter-powered car.

The energy density of a fuel is often expressed as the specific energy, that is the energy stored per unit mass (J/g), and this can be used to compare a range of different objects.

Object	Food calories per gram kcal/g ~ kWh/kg	Specific energy kJ/g	Relative to TNT
Bullet (speed of sound 300 m/s)	0.01	0.04	0.015
Battery (car – lead/acid)	0.03	0.13	0.048
Battery (laptop – Li rechargeable)	0.10	0.42	0.16
Battery (AA – alkaline)	0.15	0.60	0.23
TNT	0.65	2.7	1
Modern high explosive	1.0	4.2	1.6
Wood	3.8	16	5.9
Chocolate chip cookie	5.0	21	7.8
Methanol	5.5	23	8.5
Coal	6.0	25	9.3
Alcohol (ethanol)	6.9	29	11
Butter	6.9	29	11
Gasoline (petrol or diesel)	11	46	17
Natural gas (methane CH_4)	13	56	21
Hydrogen (gas or liquid)	26	142	53
Asteroid – 30 km/sec	100	420	156
Uranium 235	20 million	84 million	31 million

Fig. 5: Specific energy of traditional fuels and other substances. NB. The numbers in the table have been rounded. The values used in column one are: 1 food calorie (1 kcal) = 4.184 kJ, and 1 kilowatt hour = 3,600 kJ. (Source: *Physics for Future Presidents* and *CRC Handbook of Chemistry and Physics*)

Some of the energy densities in the table are initially surprising. Certainly, the fact that a speeding bullet contains a tiny fraction of the energy present in a pat of butter or a chocolate chip cookie will be of scant consolation should you be unlucky enough find yourself on the business end of one. For a moving object, velocity is all-important. Kinetic energy – that which an object possesses due to its motion – is defined as the work needed to accelerate a body of a given mass from rest to its stated velocity ($E_k = \frac{1}{2} mv^2$ where E_k is kinetic energy, m is mass and v is velocity). If the object is big and fast, the embedded energy is enormous because the mass times the velocity squared of something like an asteroid travelling at 70,000mph would be a very big number indeed. Even though a bullet travelling at the speed of sound has only 1.5 per cent the energy content of TNT, it's lethal because it deposits its small amount of energy very quickly in a small area. Indeed, despite the fact that in terms of energy, velocity is the dominant value, it's the moment of inertia that is important – when the moving object comes to a halt: when the car crashes into the tree, the asteroid hits the planet or the two hydrogen atoms collide in the heart of a star. Then, in that instant, all the energy is released.

A tonne of asteroid has 165 times the energy content of TNT. The asteroid thought to have caused the so-called K-T event (the Cretaceous–Tertiary mass extinction event that killed the dinosaurs 65.5 million years ago) was probably only 6 miles in diameter, but the collision released the equivalent of 100 tera tonnes of TNT (420 ZJ) or 1 billion times the energy of the atomic bomb dropped on Hiroshima. Asteroids can be said to have a 'significant moment of inertia' and the same is true of flywheels, albeit on a much, much smaller scale. Flywheels are rotating mechanical devices used to store rotational energy from intermittent sources, such as wind, waves or even internal combustion engines. Formula One racing cars use flywheels made of carbon fibre capable of spinning at 60,000rpm in a kinetic energy recovery system (or KERS). In KERS the flywheel recovers the moving vehicle's energy that would usually be lost during braking and allows it to be used to provide extra power during overtaking manoeuvres. Ironically, one of the world's least eco-friendly sports is teaching us how to make low-carbon transport more efficient.

The fact that moving objects have high energy densities doesn't really help us when it comes to storage. An asteroid that has come to rest has the same energy density as a stationary lump of rock and will prove just as useful at powering your car. Similarly, we can see from the table above that equally stationary lumps of metal and electrolyte – or batteries – also have incredibly low energy densities.

Batteries are stores of chemical energy. Unlike moving objects, they draw their power from a chemical reaction that takes place inside them. Despite the fact that we are currently being persuaded to run our cars using them, even the very best rechargeable batteries contain only around one-fiftieth the energy of gasoline. True, batteries are convenient, but that's the only reason to use them at all, because they really are one of the poorest energy vectors we have. First of all, a lot of the stored energy within a battery is lost as heat during the charge–discharge process – just feel how hot your mobile phone gets when you're playing a game, or simply when it's recharging. Secondly, batteries are hugely expensive. The cost is in the battery itself rather than the electricity stored within it (which typically comes from a coal-fired power station at 10 per cent of the price of petrol per mile). Batteries are filled with expensive chemicals and even re-usable varieties have to be replaced after many cycles. The true price often remains hidden. To run a small car – say a Ford Fiesta – using AAA batteries (the kind found in digital cameras and kids' toys) would cost 10,000 times as much as it would using petrol. In practice, you'd also need to fill a Transit van with these batteries, which would itself be running on diesel and travelling alongside, to provide you with enough power to drive anywhere.

Even the best rechargeable lithium batteries – the ones we find in our phones and laptops, which are about the best energy density we have at the moment – cost twice as much per mile as petrol when they are used to power electric cars. The economics of battery-powered vehicles will improve as the technology becomes more widespread, and there is a great deal of research that will lead to cheaper batteries, with higher energy density batteries and longer-lasting batteries becoming available over time. However, given the limitations of the vector, while it's likely that we will be able to choose any two from 'cheaper', 'more powerful' and 'longer-lived', it's extremely unlikely that we will be able to have all three. There are losses at every stage of energy conversion, and running battery-powered cars using our domestic electricity supply involves many more stages than we might think. The process begins at the power station with the chemical energy in the fuel (coal, gas or oil), which is converted via combustion to thermal energy, then to expansion (steam), then mechanical (turbine), and to electrical (generator), which is then transmitted down power lines, stored again as chemical energy (by recharging the battery), then finally put to use as rotary mechanical energy via a motor. The singular, big advantage of battery power over internal combustion is that in this system all of the emissions are in one place – at the power station – which gives us a much better opportunity to manage them effectively.

What makes combustible fuels so appealing by comparison is their efficiency. The really neat trick that all combustible fuels pull off (from the wood and the chocolate cookie in our table, right down to hydrogen) is that they take in oxygen from the air when the reaction occurs. The amount of oxygen consumed is dependent on the fuel. For example, one gram of hydrogen reacts with eight grams of oxygen from the air to make nine grams of water. The process is exothermic, which means we can use the released energy as a source for other applications (such as driving a motor). In this case, 142 kJ of usable energy is created:

$$2H_2(g) + O_2(g) \rightarrow 2H_2O(g)$$

TNT works in a different way, because it already contains the oxygen necessary for combustion within it. This means that the rate it can release its energy is not limited by the speed at which it can draw in oxygen from the air. TNT releases all of its energy very quickly – in an instant, in fact – hence it's very powerful (J/s), which is why we think of it as an explosive, but it's not particularly energy dense (J/g).

A chocolate chip cookie has eight times the energy density of TNT but works in the opposite way, taking an inordinately long time to release its energy as you digest it (a fact to which the expanding waistbands of North Americans and western Europeans are testament). Soak the cookie in liquid oxygen then ignite it, and you will find it behaves in much the same explosive way as TNT for much the same reason. Grinding a substance up, thereby increasing its surface area and exposure to the air, can have a similar effect. Cookies are made of flour, which is an inert substance in its normal state, but if it's airborne in high enough densities it can and does explode. This makes baking bread (and particularly milling the flour) a much more hazardous occupation than you might imagine. This sudden release of energy thanks to oxidation is the principle behind the internal combustion engine: a fine aerosol of gasoline is sprayed into the cylinder, giving each droplet the maximum exposure to the air. The resulting micro-explosions are what power the car.

Not all combustion reactions produce only water. If we take the alkanes as an example – a family of chemical compounds consisting of hydrogen and carbon such as methane, propane or pentane – burning consumes oxygen and produces heat, water and also carbon dioxide. In these reactions, carbon–hydrogen bonds are broken and carbon–oxygen and hydrogen–oxygen bonds are created:

$$CH_4(g) + 2O_2(g) \rightarrow CO_2(g) + 2H_2O(g)$$

All alkanes are chains of hydrogen and carbon molecules. As these chains get longer, the proportion of hydrogen to carbon gets smaller: 4:1 in methane, the simplest alkane, 3:1 in ethane, 8:3 in propane, 5:2 in butane, 12:5 in pentane, 9:4 in octane ... and so on, until we reach approximately 2:1 in polyethylene. The amount of oxygen consumed as a proportion hardly changes – from 80 per cent in methane to 77.5 per cent in octane – but making the water bonds creates much more energy than making the carbon dioxide bonds, so fuels that produce more water through combustion (by dint of having a greater proportion of hydrogen) have the greater energy density. One gram of methane gas produces 55 kJ with 20 per cent of the reaction mass coming from the fuel; meanwhile, burning one gram of octane (gasoline) produces only 48 kJ, requires slightly less oxygen in the reaction and results in much more carbon dioxide.

✷

The question of which is the best fuel to use is not simply a matter of energy density, but of convenience. Burning one gram of candle wax produces roughly the same amount of energy as a gram of octane or ethanol – yet candles cannot be made of methane and a car running on wax would be literally a non-starter. We need different fuels – solids, liquids and gases – for different things. Cars work best with liquid fuel[2] and there is a vast, effective infrastructure in place to ensure it gets where it's needed: into our fuel tanks.

In principle it's entirely possible to run an internal combustion engine using hydrogen instead, and there is the hugely attractive double advantage of its high energy density and zero emissions, but unfortunately that's where the good news ends. First of all, despite its appearance – a combustible gas like methane – and its abundance – the most plentiful element in the universe – it makes more sense to think of hydrogen as a vector like electricity rather than a fuel like petroleum. Hydrogen is extremely reactive, which means there are no naturally occurring sources of pure hydrogen that we can exploit: it's always locked away in molecules. True, the oceans covering our planet are essentially one-ninth hydrogen by weight, but the only means we have of breaking the H_2 from the O is electrolysis, which is a process that consumes more energy than it produces. Notwithstanding the effectively insurmountable difficulties involved in supply,

---

[2] As anyone who has taken a diesel car to a ski resort and left it outside overnight might have found out when the fuel 'waxed' (i.e. crystallised solid) in the sub-zero temperatures.

hydrogen is a terrible thing to have to store and transport. The tiniest and lightest element, hydrogen gas is prone to seepage from any container, while liquid hydrogen's boiling point is a cool –252.87°C. Swapping petroleum for hydrogen would require pressurised fuel tanks, tankers and gas stations and all the incumbent cost, engineering and safety issues. Indeed, the embedded energy of storage alone makes it essentially unusable as a transportation fuel. In short, we need a fuel with high levels of joules per gallon or litre, not as hydrogen scores well, in joules per cubic metre. For that reason alone, it's unlikely to be the fuel of the future.

Furthermore, fuel is not just 'fuel'. Not all fuel sources are equally expensive, and the market–price correlation has much more to do with convenience than with efficiency or energy density. Coal is 20 times cheaper than petroleum, which partly explains why consumption of coal is so high in the developing world. Coal however, is less convenient than oil, and no use at all if you want to run a flashlight. We could turn coal into liquid fuel easily enough – Germany did this out of necessity during the Second World War, and so did South Africa during the years of apartheid and economic sanctions. It's also quite conceivable that the market conditions will emerge where it becomes economically viable for us to do so once again, thereby enabling us to benefit from the convenience of liquid fuel, but we should pray that this day doesn't arrive, as the environmental cost would be truly catastrophic.

To succeed, alternative energy sources need to be more economically viable than coal. Taxation would make coal less appealing in developed countries, but coal usage is already in decline there and the real challenge is persuading countries gearing up their energy supply networks – China, India and Brazil – that getting their energy in the cheapest possible way is not necessarily the best possible way. Here again there are some grounds for optimism. Despite having some of the world's largest fossil fuel reserves, Brazil, the world's tenth-biggest energy consumer, produces 40 per cent of its supply from renewable sources, against a global average of just 14 per cent. We can hope Brazil's example is followed elsewhere, but as we will see, there are some thorny issues surrounding the use of sugar cane-derived ethanol to run cars. And the leaders of developing nations face many other pressures – to improve nutrition, education, health and economic competitiveness, for example – which makes the price of coal a very tempting proposition.

The list below gives a price comparison of various fuels in the UK based on 1 February 2012 prices:

Coal	3.9p per kWh
Natural gas	4.1p per kWh
LPG	10.8p per kWh
Diesel	13.2p per kWh
Unleaded petrol	13.8p per kWh
Laptop battery	£2.57 per kWh
Envira car battery[3]	£80.00 per kWh
AAA battery	£650.00 per kWh

As well as convenience, the other factors that influence cost are the energy infrastructure, which involves considerable sunk costs to establish and maintain, and the mode of delivery – pipes, pumps, power lines or packets – and the cost of cleaning up afterwards. We used to run trains using coal, and we could do that again in future, but we choose to run them on diesel or electricity instead because there's no need to worry about what to do with the ash. Cars were invented at a time when oil was cheap and supplies effectively considered endless. With abundant domestic supplies, the USA adopted the automobile as the preferred means of transport. Elsewhere in the developed world, where there was comparatively little or no oil, the uptake of the automobile was much slower. These countries – for example, most

---

[3] The most efficient battery in the market at the time of writing.

of western Europe – tended to put more investment into the development of public transport systems instead. In the US, the wealthy tend to live in the suburbs; while in Europe, the more affluent districts are in the heart of the city.

The pressures of cost, convenience and changing infrastructure should prevent any discussion about energy supply from becoming over-optimistic. Those who believe the full replacement of fossil fuels with renewables could occur in an extremely short timespan are almost certainly going to be disappointed. In 2008, former US Vice President Al Gore called for 100 per cent renewable energy in the US within ten years. Four years later and this seems foolhardy rather than ambitious, the percentage increasing from 9 per cent of fuel sources when he made the statement to just 9.45 per cent by the middle of 2011. Similarly, reports such as Jacobsen and Delucci's 2009 *A Plan for a Sustainable Future*, which aimed to get all the planet's energy from wind, water and solar power by 2030, failed to take into account the inherent inertia of energy transitions and ignored the qualitative differences between energy sources.

Both of these issues are subjected to forensic examination by Vaclav Smil, a professor in the Faculty of Environment at the University of Manitoba in Canada, in his rather sober book *Energy Transitions: History, Requirements, Prospects*. Smil's detailed analysis leads him to call for a reality check on the optimistic timings being quoted for energy transition. He arrives at the conclusion that these transitions are going to be a matter of generations rather than years and will require the implementation of policies to reduce absolute energy usage per capita in every developed country.

The scale of the coming energy transition is best illustrated by comparing the future demand for non-fossil fuels and primary electricity with the past demand for fossil energies that were needed to complete the epochal shift from biomass to coal and hydrocarbons. By the late 1890s, when the share of biomass energies slipped just below 50 per cent of the world's total primary energy supply, less than 20 ExaJoules

(EJ)[4] of additional fossil fuel supply were needed to substitute all of the remaining biomass energy consumption. By 2010 the global use of fossil energies runs at the annual rate of roughly 400 EJ, which means that the need for new non-fossil energy supply to displace coal and hydrocarbons is 20 times greater in overall energy terms than was the need for fossil energies during the 1890s.

Smil shows that, historically, energy transitions have taken a very long time. Once a fuel has achieved 5 per cent of global energy production, it usually takes many years for it to achieve a 25 per cent share of the energy market. In the case of coal, oil and gas this took 35, 40 and 55 years respectively.

> Globally, coal began to supply more than 5 per cent of all fuel energies around 1840, more than 10 per cent in the early 1850s, more than a quarter of the total by the late 1870s, and one half by the beginning of the twentieth century ...

There is nothing to suggest that later transitions will be any quicker. Indeed, given that the quantities of fuel we need to replace are many times greater, and that the more popular non-fossil energy sources suffer from constraints in conversion due to intermittency and low power densities, it's reasonable to assume that they will take much longer. A comparison between the area needed to extract, process and transport fossil fuels in the early 21st century and what would be required if we were to use bio-energy as a replacement is particularly revealing. Our entire fossil fuel infrastructure currently occupies an area the size of Belgium, some 30,000 km². The replacement biofuels network would occupy 12,000,000 km², an area equal to the US and India combined, or approximately 400 times the space required by the fossil fuels network, to produce an equivalent 12.5 TW

---

[4] Smil likes exajoules/year, which he calls EJ, whereas we prefer watts (J/s).

of energy. Our conclusion can only be that distilling bio-diesel from sugar cane and other vegetable matter really isn't the answer.

Smil is also critical of overly optimistic claims for renewable energy, including Al Gore's 100 per cent renewable electricity plan and Google's clean energy 2030 vision. For decades to come, fossil fuels will remain the primary energy source for all developed nations. There is no point hoping for a rapid technological development in the short to medium term without widespread policy implementation that addresses all of the barriers to transition.

> Difficult as it would be, reducing the energy use would be much more rewarding than deploying dubious energy conversions operating with marginal energy returns (fermentation of liquids from energy crops being an excellent example), sequestering the emissions of CO_2 (now seen as the best future choice by some industries), and making exaggerated claims for non-fossil electricity production (both in terms of their near-term contributions and eventual market shares). Or hoping for an early success of highly unconventional renewable conversions (jet stream winds, ocean thermal differences, deep geothermal). After all, a dedicated but entirely realistic pursuit of this goal could result in reductions on the order of 10 per cent of the total primary energy consumption in a single generation, an achievement whose multiple benefits could not be matched by the opposite effort to increase the overall energy use.

The bitter pill is that developed countries will need to abandon their pursuit of higher energy output in favour of greater efficiency and achieve significant per capita reductions by commerce and consumer alike. Until we create photovoltaic cells with 50 per cent efficiency, or genetically engineer algae and bacteria to produce biodiesel and kerosene on an industrial scale, we are going to be locked into a *danse maca-bre* with oil and coal. Too convenient, too cheap, too easy: the irony is that the good things about oil are what make it so dangerous, because the problem with taking generations to move away from oil and other fossil fuels is the amount of additional CO_2 we will be putting into the atmosphere in the process – and the devastating environmental impact that this will have.

Using fossilised carbon – in any form – releases CO_2, which ultimately causes global warming. The fact that we don't fully understand the implications of global warming doesn't mean that we can't be certain they will be pretty bad, simply that they will be pretty bad in ways that we haven't anticipated.

The arguments put forward by genuine climate-change sceptics like Dennis Avery in support of his contention (repeatedly trotted out by people who are paid money to have controversial opinions in newspapers and magazines) that 'Two thousand years of published human histories say that warm periods were good for people' will lead us only into a fool's paradise. There will be nothing positive about rising sea levels or an increase in the frequency of extreme weather events. Some lives may be saved as a result of milder winters, but that number will be dwarfed by those lost due to hotter summers. Economically, climate change has very little to offer beyond the promise of prolonged volatility, which markets typically abhor. Any acreage liberated for agriculture by melting polar ice-caps will be of poor quality and certainly not able to replace the arable land we expect to lose.

We do not know precisely what the implications of increasing CO_2 concentrations in the atmosphere will be. There are half a dozen climate change models regularly referred to by various organisations and bodies: each one is different. We can be sure that at least five of them will turn out to be wrong, and that the planet is on a singular path. This is the area where climate-change sceptics usually challenge. Yet the debate is somewhat superfluous, as we will never have data rigorous enough to exactly predict the climate in 40 years' time. Waiting until we have a definitive answer for that is neither practical nor desirable. It would take the best part of 40 years to discover, at which point it will certainly be too late to do anything about it.

Rather than waste energy on proving the validity of the various models, we can focus on the extensive hard data that we do have from the past. There is a robust record of global climate for the past 600,000 years contained in the Vostock ice core data, which shows that CO_2 levels have fluctuated very little during that time: between 200 and 300 parts per million for the whole period. We also know that concentrations of CO_2 will be twice as high during our lifetime, and are already higher than at any point in the past 20 million years. We also know that almost every glacier is melting. In 1928, the Upsala glacier in Argentina was the biggest in the southern hemisphere: today the ice field is a lake. In Greenland there is a vast quantity of ice, which if it continues to melt could raise the sea level by up to 20 feet (6m). The climate-change sceptics are right: we do not know for sure that increased levels of CO_2 are causing this; but if we continue to rely on fossil fuels, then we are certain to find out.

Climate change is not just about managing the temperature. It's naive to think in terms of just hotter summers and warmer winters. The acidity of the oceans is changing, 20 per cent of the world's coral is already bleached and it's not unreasonable to expect that all the reefs might completely disappear before the end of the century, along with the loss of countless species of fish. Planetary warming will also cause the Arctic permafrost to melt, which would in turn release large quantities of methane into the atmosphere. Although it will not remain there for long, this methane will act as a greenhouse gas, trapping more of the sun's heat. At the moment, the potential scale of these methane emissions is poorly understood, but again we can look to the past for what might happen, as the permafrost has melted before. The last time this happened, 230 million years ago at the end of the Permian era, annual global temperatures spiked quickly by between 4°C and 8°C, which resulted in the extinction of around 90 per cent of all species on the planet. Climate sceptics will point out that we can't be sure that this will happen, and again they are absolutely right; but if we continue to burn carbon then we are certain to find out.

The Intergovernmental Panel on Climate Change (IPCC) is the leading international body for the assessment of climate change. It has predicted that a business-as-usual scenario will see clean-burning energy steadily replace carbon fuel over the coming decades, projecting carbon intensity in 2050 of 0.45kg per watt-year. This figure is much lower than any of the fossil fuels, so it will be achieved only through a significant increase in the production of carbon-free energy. Put simply, decarbonising our energy economy is essential: not an option, a goal or an aim, but a hard and fast reality.

If we assume that on average each person needs a sustainable 2 kW of energy per year (which is roughly twice their food intake), then by 2050 we will need to produce 28 TW – an additional 11 TW on 2012 levels. We cannot afford to use carbon-based fuels to make up the shortfall, in fact somewhere between 10 and 20 TW of the total needs to be carbon-free. Even relying exclusively on natural gas, the least carbon-intensive of all the fossils, would lead to too high a concentration of CO_2 in the atmosphere. It would constitute, at best, the beginning of an irrevocable experiment with the climate, with dire consequences for us all. Fracking to access more natural gas is not the answer.

It's clear that we need a replacement for all our fossil fuels, but while oil, gas and coal are dominant, they are not the only sources of energy we use today. There are indeed many great things about oil, but the price we are paying to enjoy them is too high. The same is often said about another means of producing electricity: nuclear. Nuclear energy is a mature technology with a 60-year history. It's undoubtedly the most scaleable form of alternative energy, produces no CO_2 and generates electricity efficiently and cheaply. But it's

also the most controversial and potentially the most dangerous, and since its inception, arguments have raged about
whether the benefits outweigh the costs. Nuclear energy
undoubtedly has its rewards. We must decide whether they
are worth the risks.

10 Going Nuclear

'Our children will enjoy in their homes electrical
energy too cheap to meter.'
*Lewis L. Strauss, chairman of the Atomic Energy
Commission, September 1954*

'For the attention of the residents of Pripyat! The
City Council informs you that due to the accident
at Chernobyl Power Station in the city of Pripyat
the radioactive conditions in the vicinity are
deteriorating. The Communist Party, its officials
and the armed forces are taking necessary steps to
combat this. Nevertheless, with the view to keep
people as safe and healthy as possible, the children
being top priority, we need to temporarily evacuate
the citizens in the nearest towns of Kiev Oblast.'
*Evacuation notice issued by Pripyat City Council,
USSR, 27 April 1986*

No form of energy divides opinion like nuclear. For some,
harnessing the energy of the atom is the only way to guarantee us a future; for others, it only guarantees there will be no
future. We can all agree that it's one solution to the energy
problem, but is it the right solution?

Part of the issue with nuclear power is how we deal
with its heritage, both practically and intellectually. Nuclear
energy was developed first and foremost as a weapon of mass
destruction, and the needs of the military drove the technology for almost seven decades. The nuclear power stations
themselves, certainly up until the 1980s and arguably even
later, were primarily reactors for producing weapons-grade
plutonium – the fact that they could be used to generate
electricity was seen as a fortuitous by-product of the cooling
process (the superheated water produced by the reaction can

be turned into steam to power turbines). Furthermore, the underlying science of nuclear physics is very complicated and hard to understand, so while disingenuous communication to the public was not necessarily inevitable, it was certainly both possible and widespread. Even the politicians were bamboozled. In the 1960s, Tony Benn, then UK Energy Secretary, completely bought into the 'atoms for peace' sentiment espoused by President Eisenhower. Benn, a junior MP when Eisenhower made his speech, felt this was a case of 'beating swords into ploughshares' and became a strong advocate for the benign use of nuclear energy in the UK. He held the same view when he was appointed Minister of Technology in 1966 and thereby given responsibility for the development of the civil nuclear programme. However, his experience led him to change his mind, as he wrote in 2002:

> I was told, believed and argued publicly that civil nuclear power was cheap, safe and peaceful and it was only later that I learned that this was all untrue ... Nor are Britain's civil nuclear power stations peaceful as for many years, and still possibly today, the plutonium they produce was sent to fuel the American nuclear weapons programme, making them into – what were in effect – bomb factories. At no stage, as a minister, could I rely on being told the truth either by the industry itself, or by my own civil servants who may or may not have known it themselves.

With such an inglorious history it's little surprise that nuclear power today finds the weight of seven decades of almost entirely negative publicity bearing down on its shoulders, from the appalling scenes of carnage that followed the bombings of Nagasaki and Hiroshima, through the explosion of the reactor at Chernobyl, to the post-tsunami meltdown of the Fukushima reactor in Japan. The vocabulary of nuclear energy has become familiar. Words and phrases such as 'meltdown', 'duck and cover', 'nuclear winter', 'fall-out', 'megaton', 'dirty bomb' and 'half-life' are used and understood widely. We all know what happens when things 'go nuclear' and how dire the consequences are for everyone. Or at least, we think we do.

☀

There are many different ways to smash an atom, but at the heart of every nuclear power station is the reactor. In the reactor's core, heat is generated by a controlled nuclear fission reaction in a 'fissionable material'. Just as in conventional power plants, the conversion from nuclear to electrical energy takes place indirectly. A coolant – usually water – is pumped through the reactor under pressure to draw off the heat generated. The temperatures involved are enormous. If water is being used as a coolant, it's converted into pressurised steam and fed into a multi-stage turbine. The turbine converts the heat into mechanical energy, which can in turn be converted into electrical energy via a generator (or into kinetic energy via a submarine propeller). As the steam travels through the turbines it begins to cool and returns to liquid form inside a condenser – usually a heat-exchanger or cooling tower attached to a river – and the cycle can begin again. The process creates dangerous levels of radiation, so the reactor is surrounded by an absorbent protective shield that prevents it from seeping into the environment. The waste retains most of its energy potential and can be recycled as new fuel.

Most reactors use a moderator to reduce the velocity of the neutrons and increase the probability of fission occurring, allowing lower concentrations of fissionable material to be used than in nuclear weapons. Fissionable materials include uranium and plutonium, both of which can sustain a nuclear chain reaction. A very small amount of material can produce a huge volume of power. A single fissionable pellet the size of your little finger produces the same amount of power as burning a tonne of coal. The best moderators are materials like heavy water or graphite. These are full of atoms with light nuclei, which do not easily absorb neutrons. Reactors using either of these materials as a moderator can run using natural uranium. However, ordinary water is more commonly used instead. Unlike heavy water or graphite, light water is cheap and plentiful, but it absorbs too many neutrons to be used with anything other than U-235, a relatively rare isotope. Ordinary uranium needs to be 'enriched' – effectively it needs to be turned into U-235 – before it can be used, which adds both cost and another stage of production. The 'depleted' uranium remaining after the enrichment process is considerably less radioactive than natural uranium, but still extremely dangerous and is kept in secure storage.

During the nuclear reaction, the fissile isotopes are consumed and most of what remains is called 'nuclear fission product' – the atomic fragments left after the nuclear fissions have occurred. This is considered radioactive waste, and as fission product builds up it eventually prevents the nuclear reaction from taking place. This spent fuel is undoubtedly hazardous, but power plants produce very small amounts of waste in comparison with traditional facilities, because of the high energy density of the fuel. Some reactors do not use moderators at all, like those in nuclear bombs. Instead they use 'fast neutrons' and require much higher concentrations of fissile isotopes. However, these so-called 'breeder reactors' actually produce more fissile material than they consume.

☀

The quest to find a practical use for atomic energy began as soon as the Curies discovered that radioactive elements like uranium were releasing huge amounts of energy. Initially, the effects of radioactivity were entirely misunderstood. The fact that elements like radium appeared to produce their own energy led many to the conclusion that they must have curative or even health-giving properties. In the early part of the 20th century radium was added to toothpaste ('Doramid Radioactive Toothpaste makes teeth shine whiter'), bath salts (advertised as 'highly radioactive', bathing in them promised to 'reduce corpulence, rheumatism and gout'), and even patent medicine. Radithor was a solution of radium salts intended for ingestion that would allow users to benefit from its 'many curative properties'. Regular users died in their thousands, including the eminent industrialist Eben Byers. 'The Radium Water Worked Fine Until His Jaw Came Off', wrote the *Wall Street Journal* in 1932.

The dream of harnessing atomic energy to generate electricity was disregarded as similar quackery by many of nuclear physics' early pioneers. Even Ernest Rutherford dismissed the idea as 'moonshine', and it wasn't until the discovery of nuclear fission, just before the Second World War, that the idea gained any traction at all. James Chadwick discovered the neutron in 1932, and the particle's lack of electrical charge made it ripe for nuclear experimentation. Ten years later, a team led by the Americanised Italian physicist Enrico Fermi carried out an experiment in a Chicago squash court that was to signal the start of the nuclear age. Arranging pellets of uranium[1] and blocks of graphite into a massive, carefully planned geometric structure, Fermi

---

[1] Uranium is a silvery-white metal with the symbol U and the atomic number 92. It's weakly radioactive because all of its isotopes are unstable. The most common isotopes are U-238 (which has 146 neutrons) and U-235 (which has 143 neutrons). It occurs naturally as U-238 (99.27 per cent) and U-235 (0.72 per cent) with a very small amount of U-234 making up the remainder. Uranium decays slowly by emitting alpha particles. The half-lives of U-238 and U-235 are 4.47 billion years and 704 million years respectively (which is why they are so useful when it comes to dating the Earth and prehistoric fossils). U-235 is the only fissionable isotope that occurs naturally. U-238 is fissionable only by 'fast neutrons' (i.e. free neutrons with a kinetic energy level close to 1 MeV (100 TJ/kg), hence a speed of 14,000 km/s or higher) and can be transmuted into fissile plutonium-239 in a nuclear reactor. There are other fissile isotopes of uranium that do not occur naturally. For example, U-233 can be produced from natural thorium. While U-238 has a vanishingly small probability for spontaneous fission, in sufficient concentration both U-235 and U-233 maintain a sustained nuclear chain reaction.

constructed Chicago Pile One: the world's first nuclear reactor. On 2 December 1942, the structure became large enough to allow a nuclear reaction to occur ... producing just half a watt of electricity.

But generating electrical power was not the priority for Fermi. The aim for his team, working on what was to become known as the Manhattan Project with others led by John Manley, Robert Oppenheimer and Norris Bradbury, was to make a bomb. This military imperative remained the driving force behind the proliferation of nuclear energy for the remaining seven decades of the 20th century.

After the Second World War, the civil use of nuclear energy was widely advocated, but always under the auspices of military control. Hyman Rickover, a veteran submariner, recognised the potential advantages of nuclear over diesel as a fuel for submarines. Internal combustion engines required air delivered by a snorkel, which severely limited dive times. A nuclear-powered sub could stay submerged for years at a time. The reactor created to power this new generation of submarines was much more compact than the graphite reactors developed after Fermi. The space was saved by compromising efficient energy conversion and using ordinary 'light water'[2] as a coolant and moderator instead of the much rarer, deuterium-rich 'heavy water'. Even under high pressure, liquid water cannot be heated above 350°C, so to make up for this poor moderation, this type of pressurised light water reactor (PWR) required enriched uranium fuel, typically 3 per cent U-235. While it doesn't need to be enriched to the same extent as fuel for bombs, the system of enrichment is the same, so producing weapons-grade uranium instead is a simple switch to make. This is the reason that Iran's current claims for the civil nature of their nuclear energy programme is the subject of so much scepticism.

Light water reactors (LWRs) sparked little excitement among nuclear physicists of the early 1950s, but the US

---

[2] An abundant resource if you're in a submarine.

navy didn't need new ideas as long as it was able to exploit existing technology to meet its ends. Despite the fact that over the next ten years new research was to reveal a myriad of new ways to get heat out of a reactor, the nuclear industry in the US was locked into PWRs and all the inefficiencies inherent in their design. One of the alternative technologies discovered but never fully exploited was the breeder reactor. Rather miraculously, these both burned and produced plutonium, leading to the widely held assumption that they would be able to produce their own fuel, but no waste. The first experimental breeder reactor was built in the USA in 1951 to validate the concept, and generated enough electricity to power four light bulbs. Although several other countries have also investigated the technology, and built their own breeder reactor with equally promising results, the technology has never proved commercially attractive enough to challenge either PWRs or the later-generation LWRs: the boiled water reactor (BWR) and supercritical water reactor (SCWR).

Today, LWRs account for 359 of the world's 430 operational nuclear power plants. Of the 31 countries generating nuclear power, 27 are using this technology, and 27 new LWRs were under construction in 2012. Within them, nuclear fission occurs when a 'fissile' nucleus is split by a neutron, releasing further neutrons, which go on to split further nuclei in a chain reaction. Whether it takes place in a reactor or a warhead, the process is identical. Modern reactors produce 5 gigawatts (GW) of electricity – around a billion times more than Fermi's original experiment. Consider that this is the equivalent of six Hiroshima bombs every day and it's easy to understand why nuclear energy is such a scary business.

In 1956 Calder Hall, the world's first commercial nuclear power station, opened at Windscale on the UK's Cumbrian coast. It was followed a year later by the USA's first reactor at Shippingport in Pennsylvania. From this point onwards, nuclear capacity initially rose rapidly until it reached 300 GW in the late 1980s. Subsequent growth has been much slower, reaching just 370 GW by July 2012.

The reason for this slowdown also tells us a great deal about the timescales involved in energy transition. Plans made and decisions taken about energy today will only bear fruit several decades down the line. During the 1970s and 80s the cost of nuclear power plant construction rose dramatically, due largely to pressure group litigation and increased health and safety legislation, which extended build times significantly. During the same period the low cost of coal and gas did little to make nuclear an attractive proposition for commercial investors. A total of 63 proposed power stations were cancelled in the USA between 1975 and 1980. However, in France and Japan, both of which were heavily dependent on petroleum for electricity generation, the nuclear programme was stepped up after the oil crisis of 1973. Today, nuclear provides these two countries with 80 per cent and 30 per cent respectively of domestic electricity (although in Japan at least that figure looks set to decline due to the curtailment of nuclear energy development in the wake of Fukushima).

Opposition to nuclear power emerged during the 1960s but became progressively better organised throughout the following two decades, spurred on not only by greater public awareness but also the impact of accidents at Three Mile Island in 1979 and Chernobyl in 1986. Elsewhere, well organised opposition put paid to nuclear programmes in Ireland and Poland, while the public in Austria, Sweden and Italy voted to oppose or phase out nuclear power altogether.

In the UK, nuclear power was failing to deliver. In the 1960s it had been the cornerstone of the Wilson government's vision for a future of cheap, unlimited energy that would provide an unparalleled standard of living and the opportunity for leisure on a massive scale. Little more than ten years later, these bold promises were proving to be a pipe dream. Not only had the cost of decommissioning reactors with just a 40-year lifespan never been taken into consideration, but the problem of disposing of the waste proved to be much bigger than anyone imagined.[3] Most disastrous of all from an economic perspective was the decision to invest in second-generation technology based on home-grown designs for an advanced gas-cooled reactor (AGR) at a time when the rest of the world was getting behind the various LWR technologies. The decision to go it alone meant that if anything went wrong for the UK's nuclear programme, we were on our own. And go wrong it did. The initial project at Dungeness B in Kent was budgeted at £89 million and due for completion in 1970. By the time Reactor One was finally generating power, some thirteen years behind schedule on 13 April 1983, the cost had risen to £685 million.

Even for as enthusiastic a proponent of nuclear power as Margaret Thatcher, this price was too high. She had been attracted by the opportunity to reduce the country's reliance on coal, and in 1980 announced a ten-year programme that would result in the construction of one new reactor each year. This was subsequently scaled back to five, before the plan was abandoned completely once she became aware of the hidden costs of waste disposal and decommissioning. Where nuclear power was concerned, this lady was definitely for turning.

☀

Today, after almost 20 years in the doldrums, nuclear energy is emerging once again to stake its claim as the fuel of the future. If the predictions being made are eerily familiar – clean energy, cheap energy, safe energy, sustainable energy – then so are the concerns. Following the meltdown at three of the six BWRs at Fukushima Daiichi, after the earthquake and tsunami of 11 March 2011, Germany and Italy cancelled their nuclear power plans with immediate effect. Germany went even further, formally announcing that it would abandon

---

[3] Those six Hiroshima bombs going off each day take their toll. It's like Armageddon inside a reactor. That they survive for any time at all is testament to the quality of their construction.

nuclear energy completely within eleven years. In Japan itself, one year after the disaster, all but two of its nuclear reactors had been shut down. Yet the fact remains that while 20,000 people died as a result of the tsunami, to date not a single person has died as a result of Fukushima. And, over a quarter of a century after the worst nuclear accident of all at Chernobyl, where background radiation levels are still considered dangerously high, studies of the local fauna and flora have revealed that they show little sign of being adversely affected. Modern nuclear power plants are also undoubtedly much cleaner than their 20th-century predecessors. Indeed, coal-fired power stations actually produce more nuclear pollution, because coal contains uranium, which is released via the combustion process in the exhaust gases and concentrated in the ash.

The question for us is two-fold. Could nuclear energy possibly have a safe and sustainable role to play in the future? And if so, is it really ever going to be a viable candidate to replace fossil fuel?

Nuclear power is undoubtedly an effective means of generating electricity, but that doesn't mean it can ever be a completely safe means. There are health and safety considerations over and above the violence inherent in the process itself. LWRs of all kinds are highly stable, but every reactor is capable of overheating and exploding, even if fission is curtailed, due to the extreme temperatures and the materials involved. This is called 'afterheat' and unless it's removed by continually cooling the reactor, the fuel can quickly overheat and explode or even cause the reactor to melt down.

The buildings used to contain PWRs and BWRs are large and strong, built of steel-reinforced concrete or a thick steel shell, often with domed roofs to defend against external forces or attack. But they are not foolproof, and both PWRs and BWRs are prone to runaway nuclear fission reactions. Control of the reaction is achieved by means of control rods – typically made of boron or cadmium – inserted into the core, which absorb the neutrons that cause the nuclear fission reactions. In PWRs a high concentration of boric acid is added to the water to provide an additional control. All varieties of LWR need vast quantities of water to cool them down and prevent them from going 'critical'. This is why they are almost always built next to rivers, lakes or the sea. Of course in a nuclear-powered ship or submarine the plug can be pulled out if something goes wrong, thereby sinking the reactor into a big volume of water. Land-based reactors are not afforded the same luxury. While the fuel used is never sufficiently enriched, making it impossible for a commercial reactor to explode like a nuclear bomb, there are concerns that a combination of human and mechanical error (or an act of God, war, or terrorism) could have catastrophic effects on the surrounding people and environment. The widespread dispersal of the large amounts of radioactive product contained in the reactor would cause a direct hazard, contaminating the soil and entering the food chain. High levels of radiation can cause short-term illness and even death in humans. In the long term there is increased risk of mortality from cancer and other diseases.

Despite the utmost security measures against external forces there remain a variety of ways in which a reactor can fail. Nuclear material is inherently unstable and its behaviour unpredictable, so uncontrolled power excursions can occur. The cooling system is designed to mitigate the excess heat this produces, but in extraordinary circumstances, for example when the reactor has suffered a loss of coolant, the fuel becomes so hot that it can cause the containing vessel to overheat – resulting in a nuclear meltdown. Even when it's shut down, a reactor requires energy to power its cooling systems for a considerable period of time. The energy for this is usually provided from whatever national power grid the plant is connected to, or by emergency oil-powered generators located on site. In the case of Fukushima I, it was the failure to provide power to this emergency cooling that caused the accident. After the tsunami in Japan, the US Nuclear Regulatory Commission reported that current legislation does 'not adequately weigh the risk of a single event that would knock out electricity from the grid and from emergency generators, as a quake and tsunami recently did in Japan'.

The coastal location of many plants can be a double-edged sword, with the safety benefits of proximity to a ready source of coolant offset by the risk of flooding and tsunamis. Global warming is changing weather patterns and increasing the likelihood of cyclones, earthquakes, typhoons, hurricanes and other extreme meteorological events. The flood risk was incorrectly calculated at Fukushima and also at the Blayais plant in France, which in 1999 led to a 'Level 2 Nuclear Event' (defined as 'Significant failures in safety provisions but with no actual consequences').

If nuclear proliferation continues at anything approaching the level required to replace fossil fuel, it will place too great a demand on limited supplies of fresh water. This is likely to have a negative impact on the amount of nuclear energy we are able to produce. Using the hugely more plentiful supplies of seawater is not an alternative, due to its highly corrosive nature.

The location of new plants is also an issue. Japan, China, India and the USA all have plants in seismic zones, which require earthquake and tsunami risk to be considered in the design and build, and Fukushima highlighted the dangers of building reactors in clusters. When the accident occurred, the reactors in close proximity became involved in a parallel chain-reaction that was far more dangerous than the initial loss of reactor cooling itself. Rescue workers had to cope with three simultaneous core meltdowns and three exposed fuel pools.

Despite the protestations of various government agencies and the nuclear industry to the contrary, atomic power can never be made absolutely safe or infallible. Yet it's worth reflecting that nobody died in the accidents at Fukushima, Blayais or Three Mile Island, and while many people did die at Chernobyl, it wasn't an accident caused by malfunction or human error, but the direct result of an experiment that ran out of control. We have certainly learned from Fukushima that location is the key, but there are also other, potentially much safer ways to produce nuclear energy.

❄

Uranium is a fairly common element, approximately as plentiful as tin and about 40 times more so than silver. Unfortunately these stocks are spread evenly rather than concentrated together, which makes their recovery uneconomical in most instances. Even so, the amount of economically recoverable uranium is enough to provide us with a century's-worth of fuel at the current rate of consumption. Our current LWR technology uses this fuel inefficiently, because it's only able to function using the rare uranium-235 isotope. While recycling can make the fuel go further, there are much more efficient reactor designs using existing technology.

As we've seen, fast breeder reactors use uranium-238, which makes up about 98 per cent of natural resources. The breeder process is so efficient that it's estimated we currently have enough fuel to provide us with energy for the next 5 billion years. Although several breeder reactors have been built, the technology is not perceived to be economically viable. The price of fuel is an insignificant consideration in the overall cost of a nuclear power plant and the capital cost of a breeder reactor is over 25 per cent more than any water-cooled variant. Furthermore, reprocessing the fuel safely in a breeder reactor is hugely expensive, so despite the advantages of efficiency and minimal waste, the only breeder reactor in the world producing electricity in 2012 was the BN-600 plant in Beloyarsk, Russia.

Another alternative would be to use thorium as a fission fuel to breed uranium-233. Thorium is even more plentiful than uranium – around four times more, making it as common as lead – with 36 per cent of the recoverable reserves thought to be concentrated in India, Australia and the USA. Using thorium would increase the Earth's fissionable resources almost five-fold. Thorium-based fuels have several attractive advantages over uranium. A liquid fluoride thorium reactor (LFTR) is a development of a 1960s technology called the molten salt reactor (MSR). Research at the time showed that MSR was a safer, cleaner and more economically viable technology than all the various LWRs, but it was never adopted because it was incapable of producing the weapons-grade material demanded by the military.

Thorium reactors are forecast to cost far less than LWRs because of the simplicity of their design. An LFTR can be factory-assembled and scaled down to the point where it would be possible to transport one on the back of a truck. They are much safer, too. The core in an LFTR is unpressurised, so any increase in temperature results in a reduction in power rather than a runaway meltdown. If the reactor does overheat, a salt plug at the bottom melts away, dumping the fuel in a storage container below the reactor. End to end, the process produces 1,000 times less waste than uranium, and the vast majority of what little is produced (around 83 per cent) is safe within ten years. The remainder would require 300 years of storage, which is certainly more surmountable than the 10,000 years required for nuclear waste produced by LWRs.

In November 2012, the Nuclear Power Corporation of India (NPCIL) issued a report extolling the safety of its new Advanced Heavy Water Reactor, with construction finally set to begin on a working version that will produce electricity from India's most convenient fuel for the first time. India's nuclear programme is well into its sixth decade, and this news marks the realisation of a goal that inspired its inception: to tap into the ample thorium reserves within India's borders. Once operational, the reactor, based on a 60-year-old blueprint, could be generating 300 megawatts of power from thorium, more safely than nuclear power has ever done before, with much less environmental impact and without the risk of the kind of hydrogen explosions that tore through Fukushima. The initial reactor will be built next to one of the country's existing conventional nuclear plants, but NPCIL's technical director Shiv Abhilash has said he believes the reactors are so safe that in future they could be built right in the heart of major cities such as Delhi or Mumbai.

The technology, however, remains unproven. Thorium has for decades been repeatedly held up as a clean, safe way to produce nuclear power, but the proposed AHWR is the best that India has to show for its efforts; somewhat short of the Indian Atomic Energy Commission's 1969 prediction that domestic production from nuclear would reach 43 gigawatts at the turn of the century (in 2012 it achieved 4.8 gigawatts; just 2.3 per cent of the total electricity output). The potential of thorium is great, and interest is picking up again after the Fukushima meltdown. The biggest barrier to its emergence is institutional inertia and market economics: it's far easier and cheaper to license an existing, inefficient but proven design, regardless of its safety and environmental shortcomings, than it is to invest in an innovative technology. As such, the results of the AHWR in India will determine whether thorium can finally deliver on its promise.

The history of fission power is a lesson in the dangers of engineering compromise. Nuclear energy has been the preserve of the military, with civil objectives subordinate to weapons production. The legacy is power plants that are, at best, a trade-off. Had the primary consideration from the outset been the mitigation of risk instead, then our reactors today would be completely different and nuclear fission might actually be living up to the hype as a clean, green, renewable source of energy. With that in mind, it's probably to be expected that thorium research is being driven by China and India, two countries with expanding economies where civil necessity trumps military ambition.

Fission is of course not the only form of nuclear energy. Fusion is the power generated when two light atomic nuclei fuse together to form a heavier nucleus (the stellar nucleosynthesis we looked at in Chapter 3). It releases a huge amount of heat previously locked into the binding energy of the strong nuclear force. In our sun, fusion occurs when two superheated hydrogen[4] nuclei in plasma (the fourth state of matter) collide to form a nucleus of helium.

Fusion is the most efficient means of producing energy that we know. There is so much energy locked in by the strong nuclear force that despite the vast amount of heat produced, little actual mass – or fuel – is lost. It's the reason why the sun has continued to burn for 5 billion years, and why it will burn for 5 billion more. The fundamental science behind fusion is no more difficult to understand than fission, so perhaps it's no great surprise to learn that initial research into man-made or 'thermonuclear fusion' was initiated by the US military with a view to producing more powerful weapons. The atomic bombs dropped on Hiroshima and Nagasaki used a fission reaction to produce an explosion. In a more advanced hydrogen bomb, the energy released by the fission reaction is used instead to compress and heat the fusion fuel to begin a fusion reaction that releases huge numbers of neutrons, which in turn increase the rate of fission. The energy produced in a fusion H-bomb is 500 times greater than in a fission A-bomb.

Attempts to generate thermonuclear fusion to produce electrical energy date back to the early 1940s. The first patent relating to a fusion reactor was filed in 1946, but the optimism of early experimenters quickly dissipated once the technological and engineering challenges became more fully understood. To emulate stellar nucleosynthesis here on Earth, and produce a sufficient number of fusion reactions in plasma, we need to recreate both the massive solar gravitational pressures and temperatures that are roughly 10,000 times hotter than the centre of the sun. Making fusion requires two heavier isotopes of hydrogen: deuterium, which

---

[4] Hydrogen has three isotopes. All have one electron and a nucleus comprising one proton and either zero, one or two neutrons. These isotopes are known as hydrogen, deuterium and tritium respectively, according to their atomic mass. Helium has two isotopes. Both have two electrons and two protons and there are either one or two neutrons (He-3 and He-4).

contains a neutron as well as a proton; and tritium, which has two neutrons. In a fusion reactor these atoms fuse together to form a helium nucleus, releasing one of their neutrons, which is used to heat steam and produce electricity in exactly the same way as a traditional power station. The helium nucleus transfers its energy to the plasma, keeping it hot.

Confining this hot plasma is the biggest challenge. The high temperatures preclude direct contact with any solid material, which means it has to be confined in a vacuum. But with high temperatures come high pressures, so to prevent the plasma from immediately expanding into the vacuum an additional force is required. The necessary gravitational force is found only in the heart of stars. On Earth, confinement can be provided by a very strong magnetic field (magnetic confinement fusion) or by getting the fusion reaction to occur before the plasma starts to expand (inertial confinement fusion). Inertial confinement fusion is the process used in a hydrogen bomb. A rapid pulse of energy is applied to the surface of a fuel pellet, causing it to implode to a very high pressure and temperature. If the fuel is dense enough and hot enough, its inertia will keep it together and the fusion rate will be high enough for a small fraction of it to burn before it's dissipated. In a hydrogen bomb this pulse of energy is provided by X-rays released during the fission reaction. In a controlled fusion reactor, the driver can be a laser, ion or electron beam or a Z-pinch, which uses an electrical current in the plasma to generate a compressive magnetic field.

The attraction of fusion energy is easy to understand. The sun's longevity is testament to the fact that fusion has the potential to provide a sustainable solution to the world's energy needs: a continuous power supply with effectively limitless non-radioactive fuel (the only raw materials needed are water and lithium, and consumption of both would be extremely low) that produces no greenhouse gases, no pollutants and no long-lasting radioactive waste. According to the European Commission for Research and Innovation, to get 7 billion kilowatt hours of power a year from a coal-fired power station would require 1.5 million tonnes of fuel. To generate the same amount, a 1,000-megawatt electric fusion power station would consume just 100kg of deuterium and only 3 tonnes of lithium a year.

Or at least it would if we could get it to work.

When we were born, the reality of fusion power was estimated to be just 25 years away. It has remained 25 years away ever since. To date, the largest experiment using a magnetic confinement system has been the Joint European Torus (JET). JET produced 16.1 megawatts, which was sustained for barely half a second. JET's successor, the International Thermonuclear Experimental Reactor (ITER), is under construction in France. Designed to produce ten times more fusion power, it's scheduled for completion in 2019.

Inertial confinement has long been regarded as, at best, even more difficult to achieve, but more usually as completely infeasible. Consequently, there has been even less development of the technology. Interest has recently been rekindled in the USA and Europe, with two prototype reactors now in development: the USA's National Ignition Facility (NIF) and the EU's High Power Laser Energy Research facility. On 5 July 2012, NIF announced that its system of 192 laser beams had delivered more than 500 TW of peak power and 1.85 MJ of ultraviolet laser light to its target, and anticipated achieving fusion ignition before the end of the year. To put this in context, every pulse of power from the reactor comes from igniting fusion in a frozen pellet of deuterium and tritium, and it needs 192 pulses to simultaneously hit a pea-sized pellet more than ten times a second for fusion to occur. Optimistic reports suggested that a $4-billion pilot plant could be producing hundreds of megawatts by the early 2020s. While this is encouraging, our enthusiasm for these bold claims should be tempered by the fact that NIF is battling for funding with the $21-billion ITER. The cost of backing the wrong horse would be ruinous, so we anticipate little progress towards commercialisation of either – or neither – technology until the end of the decade.

In the case of fission, it took less than a decade to develop civil applications of the military technology. The early optimism that fusion power generation would be just as easy to achieve has proved to be misplaced. Today, more than 60 years after the first attempts to generate a controlled fusion reaction, it's still generally believed that commercial power production is unlikely to happen before the middle of the century.

＊

On balance then, does nuclear have a role to play? The answer is yes, maybe, perhaps. In its favour, nuclear is a proven technology: the only one that we can be certain would be able to provide us with the amount of power we require. However, the safety concerns have been significant enough to deter both private and public investment for the best part of three decades, which means that if we were to decide to go completely nuclear, we'd have some catching up to do.

As of February 2012 there were 439 operational nuclear power plants in 31 countries around the world, with over 60 additional reactors under construction in fourteen countries. To get near the 10 to 20 TW of output we need will require 11,000 of them. This means building one a day, every day, somewhere in the world for the next 50 years. In reality this is technically possible, but prior to anyone green-lighting that particular programme they would do well to pause and consider what happens to the risk, however small it might genuinely be today, once it's scaled up to that level across so many countries. To provide nuclear power on this scale would quickly exhaust all accessible supplies of uranium. We could use plutonium instead, which would in turn require 10,000 fast breeder reactors, each with a lifespan of just 50 years. But before anti-nuclear campaigners get too alarmed at the prospect of constructing a fast breeder reactor somewhere in the world, every other day, effectively for ever: we don't have anything like enough nuclear physicists to pull this off, nor the time to train them. In short, none of this is going to happen.

According to Ernest Moniz, Professor of Physics at the Massachusetts Institute of Technology, and head of the MIT Energy Initiative: 'There is only one reason for America to subsidise nuclear and that is the climate.' Nuclear power's zero carbon emissions will make it impossible to rule out entirely. A more realistic target of a three-fold increase in global nuclear capacity over the next 50 years would replace 700 GW of coal-fired power and reduce annual carbon emissions by 3.7 billion tonnes. For us, this is the most likely outcome. Germany is not alone in abandoning its nuclear programme, and in France and Japan it's likely to lose ground to alternatives rather than being discarded outright. Yet for many other countries, the energy security offered by power stations running on a cheap and plentiful fuel that can be easily stockpiled is likely to prove too attractive a proposition to ignore. There are also states with a long history of trouble-free nuclear energy production who do not share Germany's misgivings. For these groups, a commitment to proliferation in the short term – perhaps even under the auspices of reducing carbon emissions – is likely to have long-term benefits when the cost of fossil fuel increases, whether the result of pressures on the demand side or the supply side or both (as is most likely). Investment in nuclear power can certainly pay off in the long term: any plants built today can be reasonably expected to still be producing power at the end of the century.

Nuclear energy highlights the invidious position we find ourselves in. By 2050, we are going to require a total of 28 TW of energy. At the same time, we need to achieve an 80 per cent reduction in the use of fossils – which accounts for 87 per cent of all fuel currently – to have any reasonable chance of keeping the rise in global temperatures down to less than 2°C. The scale of this challenge is not widely understood. California has long been a pioneer in energy efficiency, but studies have concluded that even with its unparalleled levels of wealth and investment, the best it can plausibly achieve is a 60 per cent reduction. If California can't do it, what chance have the rest of us got?

We need to disabuse ourselves of the idea that one single low-cost energy source is going to replace fossil fuels within the next three to four decades. Energy is perceived as a value proposition where the only motivation is price – as anyone can testify who's been on the receiving end of a phone call asking if you'd like to reduce your fuel bills by switching electricity supplier. Electricity is a vector: no more nor less useful to consumers whether it comes from burning oil, splitting atoms, howling winds or crashing waves. Any substitution fuel is going to be comparatively expensive in the short to medium term, which means that market forces – those amoral but perennially popular instruments of change – are not really going to influence matters until we've completely exhausted all supplies of relatively cheap fossil energy. There is no way of knowing how the electricity you are using was produced; and unless we can find a way of producing it for the same price as coal, oil or gas (or less), expecting people to care enough to be willing to pay more is really expecting too much. For that reason alone, nuclear power is sure to remain in the domain of politics and government rather than economics and enterprise for the foreseeable future.

History shows us that a new technology can take a century to dominate the energy mix. From the birth of civilisation to the turn of the 20th century, the dominant fuel was biomass: wood, crop residues and charcoal. On a global scale, coal became the primary energy source only in 1900, despite having been predominant in cities such as London since 1750. Coal's position at the top of the heap lasted just 50 years before it was overtaken by the combined contribution of gas and oil. Oil itself will soon lose its prominence to natural gas, as fracking liberates further pent-up supplies. It will almost certainly take decades to get solar power up to capacity, but we will get there, because we really don't have any other options. So for the foreseeable future, nuclear fission energy remains our only viable large-scale alternative to fossil fuels, whether we like it or not. And many people do not like that proposition. They argue that the renewable forms of energy, which along with nuclear currently fuel a small proportion of our electricity supply, can be scaled up to meet the shortfall. It's certainly an attractive idea: energy that's clean, green, cheap, safe and sustainable. But can it be turned into a reality?

11 Tilting at Windmills

'Because we are now running out of gas and oil, we must prepare quickly for a third change, to strict conservation and to the use of coal and permanent renewable energy sources.'
Jimmy Carter, US President, 1977–81

A constant supply of electricity is something we take entirely for granted. Without it, modern life would be impossible. Without electricity there would be no food on the table, no clean water in the taps and no fuel in our cars. Yet despite this ubiquity, electricity is a curious thing. On the one hand it feels familiar, but it's also something of a mystery. Everyone knows what it does, but few of us can describe what it is. We can't see it, smell it or taste it – and those rare occasions when we feel it are invariably unpleasant. Humans have been aware of electricity for thousands of years – the ancient Greeks created static by rubbing pieces of amber together – but it's only in the last 150 years that we've worked out how to do something useful with it.

The national grids into which we are all connected are themselves a recent innovation. In the UK, a national supply network was achieved in 1933 from energy produced by privately-owned power stations. Electricity was used in many homes before the Second World War, but almost exclusively for lighting (there was rarely more than one socket per household, largely due to the fact that there were so few appliances available to plug into them). Building power stations was a capital-intensive business and almost all of the UK development was driven by private investment. With construction drying up completely during the war, domestic capacity in 1946 remained at the same level as in the 1930s. This placed the archaic power supply under extreme pressure as people began switching on to the convenience offered by the new electric fires. Domestic demand for electricity rose by one third in just twelve months, and this was further exacerbated by a series of bitterly cold winters. To keep the fires burning, the government placed a series of embargoes on the export of coal and introduced systematic power cuts, effectively rationing what electricity was available. The impact on British industry was catastrophic. Approximately 15 per cent of factory workers were unable to clock on, as there was no power available. Unemployment soared temporarily to 2 million.

The power supply was falling well short of the demands made by an aspirational post-war economy. At a time when people felt that things should be getting better, they appeared to be getting worse. For the prime minister, Clement Attlee, there was a clear solution: nationalisation. Existing power stations were brought into public ownership and huge amounts of public money were poured into modernising the grid. The Electricity Act of 1947 brought together production and supply for the first time in a single nationalised body. The newly formed British Electricity Authority (BEA) immediately laid down a five-year plan to build a string of power stations near their fuel supply, alongside the country's newly nationalised coal fields.

Given how much we've come to rely on electricity, it comes as something of a surprise to learn that there was a great deal of organised opposition to this initiative. The British population was far less urban than it is today and many people living outside towns and cities objected vehemently to the erection of a network of unsightly pylons that were ruining the landscape. People questioned whether it was necessary – after all, we'd survived for hundreds of years without electricity and most people were perfectly happy at the time with coal, oil and gas to provide them with heat and light. There were also concerns about the risks to public health that the high-voltage electric cables criss-crossing the countryside might cause. Questions were asked in Parliament about whether there would ever be enough demand for this electricity. Opponents argued that the national grid would be little more than an expensive white elephant, which the country could ill afford during a period of austerity. This appealed to popular sentiment, generating a groundswell of public opposition that became so fierce it risked derailing the entire project.

In response, the BEA embarked on a major public information initiative, advertising the benefits of this new, clean source of energy and giving people ideas for what they could do with it. They promised to transform the domestic world, as 'electricity brings to you machines which turn drudgery to delight'. With this new form of power, 'cooking becomes just a matter of pressing the right button'. In public information films, the dirt and squalor synonymous with the industrial revolution was juxtaposed with the cleansing energy of the electrical revolution. The delivery of electrical power became almost a moral crusade in which 'the heavy smoke clouds of the past will be dissipated for ever'. Resistance proved to be futile. Our addiction to electricity may be fairly recent, but it was also immediate.

☀

The establishment of a national grid made it possible to take a holistic view of load factors: namely, what people are using electricity for at different times of the day. A purely domestic electric lighting load means that, during the winter, there's a peak in the early morning as most people get up before dawn, and a further big peak in the evening as everybody turns their lights on again at dusk, but not a lot going on for the rest of the time. From the perspective of a large-scale electricity supply system, this is a bad thing. The big problem with electricity is that it doesn't lend itself well to storage: power is therefore generated in response to demand. Storing electricity is extremely inefficient with existing technology, making it difficult for utilities to bank energy against a time of sudden demand. This is hugely wasteful and expensive – think of how the fuel consumption of a car driven hard around town compares with the same car cruising steadily down the motorway. You can't charge customers for the masses of power you lose when you ramp down (in effect apply the brakes). Just like running a car, it's much more cost-effective to maintain the supply at a steady level throughout the day. Ideally you want people to be using electricity day and night for different purposes beyond just lighting, in order to maintain a steady supply. Furthermore, building power stations is so capital-intensive that you really do need people to be using them as much as possible to get a return on your investment. To address all these issues the electricity boards found it useful to enter the retail market and push appliances to consumers that they felt had the best load-carrying characteristics: cookers, washing machines, vacuum cleaners. Soon, electricity showrooms started to appear across the country.

The masses were plugging into the grid as soon as they could, drawn by the alluring prospect of 'labour-saving' liberation from domestic toil. By the mid-1950s there were 10 million electric irons in use. In 1954 a relaxation in the hire-purchase regulations made electrical appliances affordable to ordinary people (if not actually cheaper). Not for the last time, the public threw financial caution to the wind and started filling their houses with white goods, which one in three homes opted to pay for on the 'never-never'. Electricity had been introduced as a replacement for gas or oil lighting, but now we were using it for cooking, washing clothes, heating, cleaning and refrigeration as well. Over the years we've added televisions, radios, computers, toasters, DVD players, games consoles, kettles and mobile phone chargers. A new three-bedroom home built in the 1950s and 60s would have had around a dozen power points (including light fittings). The same house built today would have more than 60.

As people have plugged in to electricity around the world, consumption has soared. In 2000, the world consumed 13 TW of energy per year, with the US accounting for approximately 25 per cent of this amount. In 2011 this had risen to 17 TW, with 87 per cent coming from fossil fuels. Global demand is projected to reach 28 TW by 2050. We probably have enough fossil fuel resources left to meet even this demand. Moreover, fossil energy is likely to remain a relatively inexpensive fuel even in 2050 and beyond, given its adequate supply globally in its various forms. Yet even though we can in principle continue burning carbon in all its forms, that doesn't mean that we should, or should want to. The move to a sustainable, less carbon-intensive supply of energy is motivated not by the economic cost but by the environmental consequences (and their impact on us all).

An extra 11 TW is a huge amount of energy. 1 TW is enough to power about 10 billion 100-watt lightbulbs at the same time. To generate 11 TW of electricity using nuclear energy would require 11,000 nuclear power stations (more than 25 times the number in operation today). And if we are going to reduce CO_2 emissions to an acceptable level, then we are going to need enough clean or renewable energy to provide between 10 and 20 TW.

When we talk about 'renewable energy' we are usually referring to a means of generating electricity that is 'fuelled' either directly from the sun or by the planetary ecosystem. Sunlight is the primary source for wind, rain (i.e. hydro-electric) and biomass power. The core of the Earth provides us with geothermal energy, while the gravitational interaction of the Earth and the moon gives us wave power.

Our society's addiction to electricity is so ingrained, and its clean production of such importance, that one would imagine any opportunity to use fuels that are widely available, produce zero emissions and are unlikely ever to run out would be universally endorsed and adopted without question. Such is the promise of renewable energy. Yet its endorsement remains far from universal and its adoption surprisingly problematic.

There are clear parallels in the kind of opposition faced

by the original pioneers of the national grid and by proponents of renewable energy today. Deep hostility to renewables is not difficult to find. Wind farms, for example, are regularly paraded as a 'government scam' or a 'con'. So our concern cannot be only with the science of renewable energy. We must also attempt to understand why so many are dead-set against it, and why opponents find it so easy to get their views heard.

Wind power has been with us for a long time. The first records of sailing ships and windmills were more than 5,000 and 1,000 years ago respectively. The wind is effectively a secondary source of solar power, caused by rising hot air that is replaced by cooler air drawn in nearer the surface. Eagles and gliders soaring on thermals take advantage of this energy source at a local level to keep them airborne.

Currently the world's total installed wind power capacity is tiny: just 232 GW or less than 1 per cent of our annual energy needs. This looks set to change, though. From an engineering perspective it's simply a question of numbers: the more wind turbines there are, the more energy can be extracted. In the UK alone, despite the objections, the plans to generate 15 per cent of domestic energy from renewables by 2020 (and more than 50 per cent by 2050) will essentially be met by wind power.

First, the good news. There is probably enough economically extractable power available from the wind to go some way to supplying the world's current electricity needs. While there is debate about the total amount available, simulations imply that there is between 18 and 68 TW of mechanical wind power that can be extracted from the boundary layer of the atmosphere using a wind turbine.

The most common wind turbine is the mast design, beloved by green enthusiasts and loathed by climate-change doubters in equal measure. These elegant giants or blots on the landscape (depending on your view) have the turbine situated in front of the mast to cope with the turbulence. One of the key things to understand about wind is that the amount of energy that wind speed can generate is not a one-to-one function; rather energy increases by the cube of the windspeed. So if you double the speed of the wind, then you get not twice but eight times the energy. Even a small difference in the wind speed in a given area can have a big impact on the amount of energy a wind turbine can generate. It also means that the taller the tower, the greater the energy generation potential.

The flow of air over the rotor blades allows the wind turbine to extract energy from the wind by slowing it down. The theoretical maximum amount of energy that can be extracted is 59 per cent (known as the Betz limit). If the wind turbine was 100 per cent efficient, it would have effectively removed all the energy entirely, so the wind would stop. In practice, rotor efficiency is between 35 and 45 per cent, and once we have taken transmission, generation and storage into account, we are looking at usable energy of between 10 and 30 per cent of the original energy available in the wind.

Turbines are available in several sizes depending on what they are used for. The smallest 50 W turbines could be used to power a boat or caravan, while the larger domestic models, which are capable of producing 1 kilowatt, can supply a single household and also feed any excess energy they produce straight into the grid. The largest turbines can produce 7 MW. These megaliths are used in vast wind farms situated out at sea. The world's tallest, located in Germany, stands 205m high, while the largest swept-area turbine in Spain has blades 128m long. But these are exceptional. By far the most common is the 1.5 MW turbine, a staple of on-shore wind-farms around the world.

Wind turbines, whatever the size, are a remarkable feat of engineering. In Ian McEwan's 2010 novel *Solar*, Nobel Prize-winning physicist Michael Beard, head of the government's new National Centre for Renewable Energy, makes an off-the-cuff remark to get one over on a colleague and finds himself committed to a half-baked scheme to develop small-scale wind turbines that will sit on the roofs of private houses. The aim is to save the UK billions of pounds, but due to a number of unforeseen problems and mistaken assumptions the project looks more like costing billions instead. Beard's fictional misadventures aptly highlight the challenges in extracting energy from the environment. Even the 1.5 MW turbines are undeniably huge: 80m towers weigh 52 tonnes, each embedded in a 450-tonne concrete base reinforced by an additional 26 tonnes of steel. As Beard found out to his cost, the need for this robustness is due to the fact that, just like the wind, turbines are intrinsically unstable.

Critics of wind tend to focus on the main practical challenges of supply and engineering, but perhaps the biggest issue of all is that of acceptance. A lot of people just don't like the idea of covering the landscape in turbines. Sir Simon Jenkins is the chairman of the National Trust, a British charity that champions renewable energy as part of its conservation activity. In February 2012, Jenkins described wind turbines as a 'public menace' and dismissed wind as the 'least efficient' source of green energy: 'We are doing masses of renewables but wind is probably the least efficient and wrecks the countryside ... Broadly speaking, the National Trust is deeply sceptical of this form of renewable energy.'

In 2012, the National Trust was leading the fight against several proposed wind farms in the UK, including a massive off-shore project to erect 417 turbines in the Bristol Channel. Its concern is the impact that these 220m turbines will have on the coastline. There are plans to build up to 32,000 wind turbines in England and Wales over the next 40 years to meet government renewable energy targets. It's difficult to see how this will be achieved if every scheme finds itself bogged down at the planning stage.

Is the criticism fair? In terms of wind power, the key thing to remember is that just because there is an issue today, that doesn't mean that it will remain an issue tomorrow. While the concerns over inefficiency were arguably valid five years ago, there is lots of research going on all over the world making improvements, both in individual components and in the holistic generation and distribution systems.

The supply issue is obvious: wind is intermittent. The effects of storms and hurricanes can give us the impression that there is a great deal of wind energy about, but very little of it is actually usable. It isn't always there when you want it; nor can you store wind for a time when you might need it. This means that until (or unless) we make a breakthrough in energy storage, wind energy is not suitable as a base-load energy source and must be used in a mix with other renewables (or more likely, non-renewables in the short term) to meet the power demand. Denmark currently produces 21 per cent of its electricity from wind: or at least it does when the wind is blowing. In 2002 there were 54 days in the year when wind provided less than 1 per cent of the Danes' electricity – and those were the hottest summer days, during which businesses would ideally like to be running power-hungry air-conditioning systems. Because the electricity supplied is both intermittent and unsteady, the grid must be made more adaptable to deal with this, requiring further construction of buffers (capacitors) and banks of batteries.

Many of the intermittency issues can be addressed by locating turbines off-shore. Out at sea the wind is much stronger and blows more consistently. Furthermore, the off-shore electrical power potential of wind is also larger than 2 TW terrestrial potential. In some parts of the world, off-shore wind farms make geopolitical sense. In Europe, for example, wind farms can be located close off-shore, with most countries on the continent enjoying a coastline. On the downside, the engineering challenge is significantly increased, due to the sheer scale of the technology. Turbines might look fairly sedate to us, as they rotate on a distant hill-top at just 22 rpm, but the tips of their blades are moving at around 200mph. A typical horizontal farm will generally have a turbine spacing of 6–10 times the blade diameter and complex computer simulations are used to optimise the output. The speeds involved place severe constraints on their design. Wind turbines are expensive because they use an awful lot of carbon fibre composite in their blades and copper wire and rare-earth magnets in their dynamos; and they also have hugely complex rotor shafts and gearboxes.

The marine turbines are even larger. However, the appeal of off-shore sites means that they are also the focus for an enormous amount of research. Collaboration between a variety of disciplines in control systems, electrical engineering, energy storage technologies and mathematics is bearing fruit in the development and demonstration of next-generation wind power systems. Investment in wind-tunnel test facilities is helping control engineers to develop a better understanding of the aerodynamics involved, allowing vibration analysis and control, and the modelling and simulation of complex structures and interactions such as vortex shedding, which is a major cause of instability. Electrical engineers have developed a unique magnetic force/torque transmission system that improves reliability, reduces maintenance, produces more energy and is completely immune to the kind of wear and tear experienced by mechanical gears and eliminates the problems of gearbox-related failures altogether. Applied mathematicians have developed and are continuing to improve high-frequency radar systems that can measure surface currents and wave direction to provide accurate wind measurement for the purposes of planning, installation and operation. SMART systems for health monitoring and mitigation of the materials used in turbine construction are in development. Long-distance health monitoring will drastically reduce maintenance costs, by identifying problems as they arise and minimising the risk of breakdown. Structures are also being designed to be self-healing while maintaining their integrity, which will similarly reduce costs across the board.

The rapid growth in the installed wind capacity and the continuing improvements make us very hopeful for the future, but, like nuclear, it would be wrong to think that wind can blow away all of our energy problems. Even if we can be confident that the considerable engineering challenges will prove to be surmountable, there would remain the issue of transmission loss by the time energy from off-shore reaches the land, where it's needed. This issue is particularly salient in the US, where most regions are remote from the coast: off-shore here is not an option. Moreover, to generate energy in the range of 10–20 TW would involve vast wind farms located way out at sea, but just because there is 10–20 TW of extractable energy does not mean that we would want to extract it. While a single turbine has no effect on the global atmosphere, building the 4 million or so required to deliver 10 TW would undoubtedly interfere with atmospheric circulation, diminishing the efficiency of extraction on a huge scale. It's not clear what the impact of removing 50 per cent of the energy from the atmosphere will have on the weather, but we can be fairly sure it will be considerable. Adding that all up over the entire globe, and taking intermittency and practical siting constraints into account, produces a total electrical power potential of about 2 TW for terrestrial wind power, with the same again from off-shore. Good, a huge increase on today's levels, but still around 80 per cent short of our target.

Efficiency will continue to improve and maintenance will get easier, although it's difficult to see capital costs reducing much further, as most of the economies of scale in turbine manufacture have already been made. We should all be enthusiastic about the prospect of more and more capacity being installed and integrated into grids, perhaps as much as 4 TW, but while we can address the issue of intermittency, we can never make it go away completely – and for that reason alone, wind power is destined to remain forever problematic. It's a similar story with all the renewables. While there's much to be positive about in all the various forms, there are always too many caveats and downsides that prevent us getting really excited about them. Invariably they neither live up to the promises of their proponents nor down to the castigations of their opponents.

Wind is the only large-scale green energy ready for market, but many policymakers and pressure groups, including the National Trust, believe that the UK's clean energy targets will be achieved by investing in the more nascent technologies involving water power and biomass. On the face of it water seems like a mature form of power. More than half of the current installed capacity of renewable energy is made up of large hydro-electric systems. The next most important contributor – at about one sixth – is biomass (mainly wood for domestic heating) followed by solar thermal (again, predominantly for domestic use). Like nuclear, biomass and water's newsworthiness is in inverse proportion to its contribution. The combined energy they produce currently is barely significant on the global scale, so is there any evidence there to support the view that they have this great potential?

Like wind, hydro-electricity is powered by sunlight. Water is evaporated from the seas by the sun, it falls as rain on the land and flows back to the sea through rivers and streams under the effect of gravity. Kinetic energy is extracted from the water as it makes its way back to the sea and converted into electricity. There are many, usually very large, hydro-electric schemes in the world today. Their purpose is usually two-fold: as well as producing electricity, they also manage water supply downstream for sanitation, drinking and irrigation purposes. The Hoover Dam on the Colorado River is prototypical of this kind of scheme: a massive civil engineering project undertaken in the 1930s and a wonder of the modern world. 7.5 million tonnes of concrete (that's 3.2 km^3) was used in its construction. It took five years to build and cost 112 lives. The Hoover Dam has a power capacity of 2 GW and has produced an average of 4.2 TWh (terawatt hours) per year since its completion in 1936. The primary mode of the operation is not to generate electricity, but to control the supply of water downstream to 400,000 ha of prime agricultural land. The electricity is generated when the water is flowing and is fed into the national grid, where other power sources are managed to accommodate the supply. The money generated from the energy supplied has more than covered the costs of constructing the dam and its upkeep. The project has been in energy surplus (i.e. producing more than it cost) for more than 50 years. Hydro-electric's greatest strength is in pumped-storage. It's effectively an inexhaustible sink of renewable energy. At times of excess supply, water is pumped to an upper reservoir as a store of potential energy, and during periods of heavy demand it creates power by flowing through turbines to a lower reservoir.

So far so good, but hydro-electric projects are always mired in controversy, and more often than not they cause many more problems than they solve. It's an application that needs to be large and grid-based to achieve the necessary economies of scale. Consequently the environmental impact is enormous: dwarfing the financial cost. Every hydro-electric project comes with a unique set of problems, because the human and physical geography is unique to each scheme.

The Three Gorges Dam on the Yangtze River in China is the biggest installed hydro-electric scheme in the world. It has a capacity of 22.5 GW, which is used to generate about 80 TWh/year. In terms of production it's second only to the Itaipu Dam on the border of Brazil and Paraguay, which generates 90 TWh/year from its 14 GW capacity. The construction of Three Gorges caused untold environmental and cultural damage, displacing 1.3 million people and destroying unique ecosystems, causing the loss of many rare species. The economic benefits are clear: it has increased the rivers' shipping capacity and produces lots of electricity without recurrent carbon emissions, but the cost has been irreversible ecological damage and the increased risk of landslides for millions of citizens.

While far fewer people were displaced during the construction of the Itaipu scheme in South America, the world's largest waterfall at Guaira was first submerged, and then demolished to facilitate safer navigation. In 2008, Itaipu supplied 90 per cent of Paraguay's electricity and 15 per cent of Brazil's. However, since construction began in 1970 it has been a constant source of conflict between both countries and also with Argentina, due to its effect on the watershed. Argentina was involved in agreements to control the amount of power generated by the downstream flow of water, but at the time of construction all three governments were military dictatorships. Argentina's primary concern was that hostility between the nations could lead to the dam being used as a weapon to flood Buenos Aires. Again, despite the political and environmental issues, when viewed in purely economic terms, the dam has been a success. Although Itaipu cost nearly $20 billion to build over a fourteen-year period, it's already in energy surplus and will remain so indefinitely.

It is the long-term cost-effectiveness of hydro-electricity schemes, along with the huge amount of capital required to construct the dams themselves, that makes it tempting for the cash-rich and power-hungry to exploit poorer states with hydro-electric potential. Completed in 2006, the Akosombo Dam on the Volta River in Ghana is an example of this form of neocolonialism. Its construction resulted in the formation of the world's biggest man-made lake. At some 8,500 km² and covering 3.6 per cent of Ghana's land, Lake Volta was created not for the benefit of Ghanaian citizens, but to provide electricity for the then American-owned VALCO Aluminium Company. Around 80,000 people were forcibly relocated during its construction and some of the best agricultural land in the country was flooded, but the greatest damage has been wrought on the health of the local population. Water-borne diseases are at epidemic levels, because the weeds in the newly formed lake provide a perfect breeding ground for the swarms of black-flies and mosquitoes that are carriers of bilharzia and malaria, and a species of snail that causes river blindness. Akosombo initially proved a great stimulus to the Ghanaian economy, but due to a growing population, more and more of the dam's energy has been diverted from aluminium production and it's now insufficient to cope with domestic power demands.

Smaller-scale hydro-electric facilities do exist. There is

getting on for 100 GW of small-scale hydro-electric infra-structure already installed around the world, with 70 per cent of this to be found in China as part of its village electri-fication project. Run-of-the-river systems are comparatively modest operations that use river flows with no lake storage at all and can generate power up to 10 MW. These systems usu-ally serve isolated communities and while in principle they could be connected to the grid, they very rarely are in prac-tice. However, with minimal need for civil engineering work involved, they have less environmental impact than their grandiose cousins and provide similar high levels of return on investment.

There is undoubtedly further demand for untapped hydro-electric capacity around the world. Collectively, the power available is significant. The total hydrological energy potential of the Earth – the water flowing in every river, lake and stream – is approximately 4.6 TW. Clearly, it's neither possible nor practical to extract all that energy: this would require damming up every one of those rivers, lakes and streams. Consequently, the total extractable is estimated at roughly 1.5 TW. At that level, we might even be able to quadruple the current amounts, especially if we follow the small hydro-electric model, but we also have to factor in the costs in terms of the habitable and agricultural lands lost to large-scale hydro-electric facilities as well as the ecological and environmental damage they cause. Yet even if we do achieve this, and combine it with 4 TW from terrestrial and off-shore wind, it will come nowhere near the requirement for an add-itional 10–20 TW of carbon-free power. Hydro-electric can grow – and it will grow – but the price we pay may well end up being too high and its impact will remain small.

✺

A much less developed form of hydro-energy is wave power. Waves are generated by wind passing over the surface of the sea and by tides caused by the moon's gravitational pull. There is realistically around 2 TW of extractable wave energy, with most potential coming from the western seaboard of Europe, the northern coast of the UK, and the Pacific coast-lines of North and South America, southern Africa, Australia, and New Zealand. The challenges involved are even more significant than with off-shore wind, and consequently the main stumbling block is cost. Seawater is highly corrosive and the oceans are very hostile environments. Power plants must be robust enough to withstand the-harshest condi-tions, yet able to produce a reasonable amount of power. Inherent safety mechanisms mean that efficiency drops dur-ing rough weather. As with wind power, there is also the question of intermittency and transmission, making it suit-able only for supplying coastal regions. Research also sug-gests that the impact of wave farms on the environment is significant (although this should be considered alongside the environmental benefits of a reduction in the use of fossil fuels – something that is never brought up by enthusiastic global-warmers). The current installed capacity is tiny, but with investment there are great opportunities.

The most optimistic estimates for hydro-electricity, wave and wind provide us with a combined 7.5 TW of energy. Even if we assume that the issues of transmission and inter-mittency are overcome, that still leaves us some way short of the target. It's unlikely that biomass can make up the short-fall. Biomass is energy captured from the sun by plants via photosynthesis. Yes, it's carbon-neutral and renewable, but it's also hugely inefficient. Under optimal conditions, during peak periods of sunshine, less than 1 per cent of the recov-erable sunlight that plants receive is stored as energy. To produce 20 TW of energy we would need to cover 31 per cent of the Earth's total land area in farms devoted exclu-sively to the production of biomass. The total amount of land with crop production potential is estimated at 2.45×10^{13} m^2, with 0.897×10^{13} m^2, or just over one third of that, currently under cultivation. Additional land will be required to feed a population of 9 billion people. Assuming that we continue to improve agricultural efficiency and crop yields (a contentious assumption in itself) an additional 4 million square kilometres will be required by 2050, leaving us with 13 million square kilometres for biomass production. Today, under the most favourable circumstances, we can achieve a yield of 8.5 tonnes of oven-dry biomass per hectare, so if this entire area was covered in biomass farms (that's 10 per cent of the Earth's total land; in effect, every square metre of cultivatable ground not used for food production), the energy produced would be 7–12 TW. The higher figure is based on breakthroughs in biotechnology that allow us to double the yield to 15 oven-dry tonnes per hectare per year. Many people believe that achieving an 8.5-tonne yield is too optimistic. Then there are the challenges of transporting and processing this much biomass. However, all these arguments are somewhat academic, as there isn't enough fresh water to support agriculture on such a scale anyway.

Rather than wrestle with these impractical theoretical quandaries, it seems clear that the challenge to science is not to improve the efficiencies of an ultimately flawed technology, but to research and develop second- and third-generation solutions. Improving the efficiency of photosynthesis is an obvious goal. Increasing efficiency by a few per cent would drastically increase yields and reduce the amount of land required for cultivation. Likewise, developing more efficient means of converting cellulose to ethanol or bio-engineering ethanol fuel-producing cells would enable biomass to make a significant contribution to the 10–20 TW of energy we require. But even making the most optimistic assumptions, it's clear that biomass cannot be relied on to meet the demand by itself.

☀

Our first personal experience of geothermal energy recovery was in the city of Beppu, on Japan's Kyushu island. A series of steaming chimney stacks acted as odorous milestones for our journey to the hotel, filling the air with the bitter scent of sulphur and brimstone. Beppu sits in a volcanic region where superheated water from deep within the Earth escapes to the surface to form hot springs. These feed the many 'onsen' or public baths, which are a great tourist attraction throughout the Kyushu region. They also supply much of the domestic heat and generate electricity as well. This type of district heating has a long but narrow history. The earliest scheme to use a hot spring to provide domestic heat was in Boise, Idaho, USA in 1892. Hot water was pumped through a series of pipes like a community-wide version of central heating. The first plant to commercially generate electricity using geothermal energy was built in Laderello, Italy in 1911. It operated on the same principle as traditional power stations, but without the requirement for a fuel source to superheat the water to drive the turbines. Despite its success, it remained the only commercial system in the world until an installation opened at Rotorua, New Zealand in 1958.

The process of extracting geothermal energy from hot water springs is well understood, and the development of the plastic (polybutylene) pipe has opened up many additional opportunities to extract heat from beneath the ground. Unfortunately, there's very little recoverable energy worldwide: about 10 GW in total, less than India's installed wind capacity. More complicated is the process of extracting geothermal energy directly from Earth's core. In volcanic regions, the heat and energy contained beneath the Earth's crust is relatively close to the surface and, in principle, it can be used to heat water to provide electrical energy. In Russia, a binary cycle system has been developed that uses these relatively cooler geothermal sources together with a heat exchanger, to remove warm water from the ground and send colder water back down in its place. Advances in the efficiency of binary cycles mean that the technology is becoming more widely available, but they can only be used at the edges of tectonic plates – the places where geothermal energy is close to or at the surface. These are some of the most geologically unstable places in the world: prone to tremors, earthquakes and volcanic eruptions. The civil engineering challenges are considerable, and intervention can have negative consequences. In December 2009, a binary geothermal project was cancelled in Basel, Switzerland after it caused several minor tremors of up to 4.5 on the Richter Scale, damaging many buildings. A study commissioned by the canton concluded that Basel was 'unfavourable' for geothermal power generation.

In some parts of the world geothermal energy provides a cost-effective and environmentally friendly source of energy. Iceland generates around 400 MW of electricity from geothermal sources, about a third of its needs, and nearly all domestic heating is produced this way, but it's a very small country with a population of just 318,000 (as anyone who invested in one of its banks during the early part of the 21st century will be able to testify). There is still plenty of potential and that will surely be exploited, but geothermal, more than any other source of renewable energy, has a long history of seemingly good ideas that fail to work in practice.

☀

We can be cautiously enthusiastic about all these forms of renewable energy, but there are no panaceas here. Even with the utmost goodwill and accepting the most optimistic projections, we must conclude that we will not solve our energy problems with these technologies alone. Indeed, the best projections we can feel confident about fall massively short of the total requirement today, let alone for 2050.

We have to accept that it's going to be much easier to use less energy in the future than it will be to generate more power. This is why the UK government has a long-term project of energy-saving in domestic buildings: lighting and heat comprise a big part of our energy budget, and insulation and low-energy lighting are relatively cheap. That's not to say there isn't a considerable role for renewables to play. Each technology has its shortcomings, but in all cases these are outweighed by the potential. The fact that the technology is, in almost all cases, at the very early stages of development, presents critics with a soft target. We should be suspicious of the media's willingness to provide them with a platform. Coal, oil and gas have been subjected to 350 years of research and development – is it really any surprise that wind, wave or biomass fail to stack up economically alongside them at the moment? What we need is the investment to develop these alternatives quickly. Time is of the essence.

There does, however, remain one final source of energy to explore. It's the primary source of energy for everything that we've discussed in this book, the one with the most potential, the one that's most scaleable, the one that will never run out – yet paradoxically the one that is most under-exploited.

12 Shine

'Any intelligent civilisation on any planet will eventually have to use the energy of its parent star, exclusively.'
Carl Sagan, cosmologist

Our sun is the source of all the energy on the planet. From the winds and waves to the fuel in our cars or the food on our plates: it's all nothing more than converted sunshine in different forms.

The history of solar power dates back much further than oil, gas or coal. The ancients understood the relationship between sunlight and biomass – the wood, animal dung or peat that they burned to provide heat and light – even if they didn't understand the process. They knew that sunshine made the crops grow, but they didn't know how it did this, and their ignorance led them to the conclusion that its powers were supernatural.

The sun itself was worshipped as a god by relatively sophisticated civilisations around the world: from the Incas and the Aztecs in North and South America, to the Babylonians, Persians, early Hindus and Egyptians in Asia and Africa. In Europe the sun provided the Greeks with not one but two major deities: Apollo and Helios, to whom they devoted many magnificent temples. The Greeks' interest in the sun went beyond blind worship. As early as 400 BCE they were the first people, as far as we know, who attempted to harness its power by employing passive solar design in their architecture and town planning. Greek cities were constructed in a manner that would ensure each building enjoyed the optimum benefits of direct sunlight. These designs, like so many Greek ideas, were later adopted and improved upon by the Romans, who used glazed windows to trap solar warmth, eventually constructing the first greenhouses to nurture the exotic plants they had collected from across their empire.

Solar energy was on occasion put to even more impressive uses. Writing during the second century CE, the Assyrian satirist Lucian recounts a remarkable tale of solar power being used as a weapon during the siege of Syracuse, which had taken place some 350 years earlier. According to Lucian, the great mathematician and inventor Archimedes (he of the 'Eureka!' moment and water screw) destroyed the Roman fleet with fire directed by 'burning glasses' mounted on the ramparts of the city. The weapon appears to have been an array of highly polished bronze or copper focusing mirrors that were used to direct concentrated sunlight onto approaching ships with such intensity that they burst into flame. The same burning glasses are mentioned in another account written a few hundred years later by Anthemius of Tralles.

Archimedes' heat-ray has been the subject of debate ever since. Renaissance philosopher René Descartes rejected it as a myth, but modern researchers, using only methods and materials that would have been available to Archimedes, have managed to get a mock-up of a Roman warship to burst into flames. In this experiment a plywood ship was coated in tar, which undoubtedly helped to achieve combustion, this kind of coating being widely used to waterproof hulls around the time of the siege. Whether the heat-ray is fact or fiction we'll never know for sure, but the tale itself highlights just how long we have known about the sun's potential. And it's a potential that remains hugely under-exploited some 2,000 years later.

※

We are considering solar power separately from the other renewables for good reason. Not only is there ample solar energy to meet our needs, but it is in fact the only renewable resource with the capacity to satisfy our global, carbon-free target of 10–20 TW by 2050. Each year the Earth receives 174 petawatts (PW) of solar radiation. Just under a third of this is reflected back into space, while the rest is absorbed by the oceans and land, and by clouds in the atmosphere. There is 89 PW of solar energy soaked up by the Earth annually, and considering all the practicalities involved, the total available to us through on-shore power generation is about 600 TW. How much of that we end up with depends on the fraction of land we can devote exclusively to the purpose of capturing this energy and converting it into electricity. Even the most conservative estimates put the figure at 50 TW, but many believe that well over 1,000 TW could be harvested. This means that if solar farms could be made to run at 10 per cent efficiency, then at least 60 TW of power could be supplied from terrestrial solar energy.

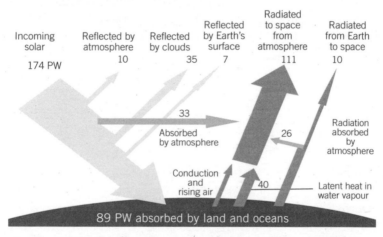

Fig. 6: The Earth's solar radiation budget. (Source: NASA, Vaclav Smil and Frank van Mierlo (Wikipedia Commons))

The amount of power available to us from direct sunlight is, by any measure, a huge number. By way of comparison, the total output of global plant photosynthesis, which is what keeps the biosphere running, is just 90 TW of energy. In theory there is almost no upper limit to how much energy we could capture. A Dyson sphere is a hypothetical megastructure for harvesting solar energy on a cosmic scale, a concept proposed by the theoretical physicist Freeman Dyson. It is essentially a massive system of orbiting solar satellites that entirely encompass a star to capture almost all of its energy. Dyson speculated that this kind of structure would prove a logical consequence of the escalating energy needs of all technologically advanced civilisations. He even suggested that searching for evidence of these artefacts could be the most effective way of detecting advanced intelligent alien life-forms.

Techniques for collecting solar energy date back to ancient times using a variety of methods, but today we can broadly categorise them all depending on whether they passively or actively capture and convert solar energy. Passive solar capture includes things like orientating buildings so that they receive the optimum amount of sunlight, or constructing them with materials that absorb heat or that possess light-dispersal properties, and designing interiors so that the air is circulated naturally. There are also a number of methods for actively converting sunlight into stored energy. Biomass produced through photosynthesis, which we discussed in the previous chapter, is one. However, natural photosynthesis is pretty inefficient, requires vast areas of land per unit of energy produced, and certainly isn't going to get us any more than a few fractions of a per cent of what's available.

Another way, which we are all becoming familiar with, is to use photoelectric or photovoltaic (PV) panels and solar thermal collectors to harvest the energy. Solar panels work by exploiting the photoelectric effect and the photovoltaic effect, two closely related but fundamentally different processes. Some materials, known as photodetectors, are sensitive to light and other forms of electromagnetic energy. When certain photodetectors – various forms of silicon or compounds such as cadmium telluride, for example – are exposed to light, a voltage or electric current is created. This is the photoelectric effect and exposure to any sort of radiation will produce it.

Light is composed of tiny packets of energy called photons. Each photon contains a different amount of energy depending on its wavelength, which is what gives the light spectrum its array of different colours. The photoelectric effect occurs when photons are absorbed from the visible and ultraviolet parts of the spectrum.

When photons strike the material in a photovoltaic panel, some are reflected, some pass right through, but some are absorbed. It's these absorbed photons that produce electricity. They transfer their energy to the negatively-charged electrons orbiting the nucleus of the atoms of the semiconductor. This allows the electrons to escape from their natural position, leaving a 'hole' behind them, which behaves like a positively-charged particle. These liberated electrons and holes head off to separate electrodes, generating a voltage. The photovoltaic effect differs from the photoelectric effect because the electrons are not emitted but transferred between different quantum bands in the material. Together, these escaping electrons form an electric current, which can in turn be used to create a build-up of voltage between two electrodes and then driven through an external load. Devices using the photovoltaic effect are usually called solar cells. Illuminating the cells creates an electric current as escaping electrons and the holes they leave behind are swept in opposite directions, because the depleted semiconductor junction now has its own built-in electric field.

If this sounds complicated, then that's because it is. The photovoltaic effect was first observed by Alexandre-Edmond Becquerel in 1839, but the underlying process wasn't fully explained until 1905 when Einstein worked out how the photoelectric effect operated. The relationship between light and electricity was widely understood in the 18th century, when it was discovered that light could produce an electric charge – a property known as electromagnetic radiation – but working out the fundamental details took some time. Einstein's paper 'On a Heuristic Viewpoint Concerning the Production and Transformation of Light' contained a simple description of 'light quanta', or photons, and used this to explain the photoelectric effect and associated phenomena. By his standards it was a relatively straightforward explanation: the premise that quanta of light are discretely absorbed ultimately leads to all the features of the phenomenon and the characteristic frequencies. Even so, it wasn't until the arrival of quantum mechanics and the establishment of the principle of particle–wave duality that photovoltaics evolved into a field of research rather than just a term. It should come as little surprise that Einstein was awarded a Nobel Prize for this paper in 1921: he had effectively cleared up a 60-year-old mystery.[1]

The modern field of photovoltaics is devoted to the practical application of this physical phenomenon: literally producing energy from light. This is usually achieved through the use of cells made of a photovoltaic material (typically pure crystals of silicon sliced into wafers) that is capable of absorbing light to generate the electrons and the holes they leave behind in pairs or, in some cases, a quantum variant of the two in a bound state called an 'exciton'. These various types of charge-carriers are first separated out and then extracted to an external circuit.

Thankfully there's also a much simpler way to understand how a photovoltaic cell works: just think of it as an LED bicycle lamp running in reverse. In a bike lamp the charge is stored in a battery, which provides a current of charged electrons. When you run this electric current through an LED, what's actually happening is that these electrons become excited and briefly jump up a quantum level. When they jump back down, the pent-up excitement is released as a photon of light. A photovoltaic cell does exactly the opposite. It absorbs the photons of light at various wavelengths that excite the electrons in the material, allowing them to escape. Electrons are negatively-charged so the holes they leave behind in their parent atoms become positively-charged. This charge

---

[1] You may question why Einstein was awarded the Nobel Prize for this discovery and not for his much greater, crowning achievement, the Theory of Relativity. Certainly the timing was right – Eddington had recently confirmed Einstein's predictions of a curved space–time continuum, by observation of a solar eclipse – but such are the Byzantine politics and vagaries of the prize committee!

separation can now be collected as current. You could attach a photovoltaic cell to the top of your bicycle lamp and use it to recharge the battery.[2] In effect this is what happens if you use a solar-powered calculator (like the one employed to check all the calculations in this book) indoors and after dark, because it's using the artificial light from fluorescent lamps as its energy source.

There are several photovoltaic materials, but each one works only with a narrow range of the spectrum, and the intensity of the light source has a huge bearing on the current output. The first truly viable photovoltaic cell was developed at Bell Laboratories, New Jersey in 1954. Calvin Souther Fuller and Gerald Pearson developed a system using silicon (diffusion-doped with boron) that achieved 6 per cent efficiency (in terms of converting the energy of sunlight into electricity). It might not sound like much, but previously the best cells, using selenium rather than silicon, had struggled to achieve 0.5 per cent. The first commercial applications were fairly humble in ambition, essentially little more than a gimmick for children's toys, because the cost of the electricity generated was extremely prohibitive: around $250 per watt in even the brightest sunlight at a time when coal-fired plants were operating at as little as $2 per watt.

☀

Our energy requirements are still some way short of requiring a Dyson sphere, but solar technology was promoted from being little more than a curiosity by a similarly cosmic application: providing a power source for the burgeoning space exploration industry. In 1958, US scientists were

---

[2] The question you might be tempted to ask is: why can't you put a PV cell in front of the lamp to charge the battery? The answer is, because you would get less energy out of the PV cell than you put into the LED from the battery, so eventually it would run down. In effect, what you'd be trying (and failing) to set up is a perpetual motion machine. Perhaps most importantly, very little light would escape, thereby rendering the lamp useless.

searching for ways to extend the life-cycle of satellites. They were incredibly expensive to launch, but due to their limited battery power they were operational for a limited period of time, so they seemed to offer little in the way of practical applications. The original and much vaunted Sputnik 1, launched the previous year, had in reality been little more than a metal orb the size of a football that beeped at regular intervals until its battery ran flat after 22 days in orbit. It was proposed that solar panels could be a cost-effective means of extending mission times, which would require no major changes to the design of either the satellite or its power systems. Despite initial scepticism, solar panels proved to be a huge and almost immediate success. Vanguard One, the fourth artificial satellite to be launched and the first to be solar-powered, was still communicating with the Earth in 1964, six years after lift-off.

Despite this and further high-profile successes, most notably the Telstar communications satellite launched in 1963, improvements in design were painfully slow. For almost two decades solar power remained an exclusively space-age technology. Ironically, its success in such an advanced field was the reason for its lack of progress elsewhere. Solar's power-to-weight ratio was higher than any rival technology, so space agencies were willing to pay top dollars and roubles for the best possible cells. While the space race was going flat-out, demand was high and price no object, so there was very little incentive for manufacturers to invest in the speculative development of less efficient technology for unproven industrial or domestic markets. When the price of solar technology did come down in the late 1960s it had nothing to do with economies of scale, but was rather a consequence of improvements in the production of silicon (in wafer form, the raw material for the cells themselves) by the semiconductor industry. The demand for transistors by the nascent computer industry led to the increased availability of large single crystals of silicon ('boules') at much lower prices. But at $100 per watt the cost of producing electricity still remained prohibitively expensive for most people. For the terrestrial market to emerge the price also had to come down to earth, but with the rich pickings from the space race providing companies with little incentive to reduce margins, an impasse was reached.

A new, cheaper method for producing high-quality silicon became the catalyst for mass production. Material scientist Elliot Berman had been refining a process for producing silicon in ribbons since the early 1960s without attracting either interest or funding. Quite why he spent so much time on something that looked like a product without a market is anyone's guess. In 1969 he was introduced quite by chance to a research team at the oil company Exxon. They were looking into a variety of energy technologies that could be developed for the 21st century. The group predicted that the cost of electricity would increase considerably over the next 30 years and concluded that the market conditions at that time would make alternative energy sources commercially viable. They were particularly interested in solar power, and Berman's work with the semiconductor resonated strongly with them. Almost immediately they decided to join forces and renamed themselves the Solar Power Corporation (SPC).

SPC's first action was to identify the potential market opportunity for a new solar energy product. They determined that reducing the price-point to $20 per watt would generate significant demand and, on the understanding that the ribbon technology was years away from achieving this, evaluated all the available materials to see if any of them might help to hit this price-point.

The solar cells used on spacecraft were based on the standard semiconductor production process. They were in effect the same silicon wafers used by the emerging computer industry to make transistors. For the purpose of generating solar power they were vastly over-engineered. A big breakthrough occurred when Berman realised that silicon from rejected boules, which was considered sub-standard for making transistors, would be perfectly adequate for solar cells. By making panels using the electronics industry's waste materials instead of its premium product, his team was able to achieve a five-fold reduction in price in just a few years. By 1973 SPC was in business: commercially producing panels at $10 per watt and selling them at $20 per watt.

Despite making the breakthrough in price that they believed would open up the market, like so many innovators before them, the team at SPC encountered a number of unforeseen and, to them, rather curious barriers to trade. Like all start-up entrepreneurs they looked for the so-called 'low-hanging fruit' and identified the navigational buoy market as a 'quick win'. Buoys were powered by batteries, which were expensive to purchase, install and replace. Surely any fool would recognise the enormous comparative benefits offered by endless solar power instead? After approaching several manufacturers, like all the most successful entrepreneurs they immediately stopped using glib phrases like 'low-hanging fruit' and 'quick win', because it was clear that despite the obvious benefits, no one was buying.

The issue was that the market leader in the sector, a business called Automatic Power, was also a battery manufacturer. For them, solar was a threat that could eat into their battery profits, so they opposed it at every step, putting pressure on their customers to do the same. Their commitment to protectionism was fierce. Automatic Power had already purchased a solar-powered navigation prototype from another business, which it had immediately shut down and shelved. Market entry appeared to have been blocked, but this very-difficult-to-reach fruit was eventually plucked thanks to a partnership with a rival battery manufacturer, founded by ex-managers of Automatic Power. One could assume that it might not have been just the commercial opportunities that attracted Tideland Signal, but also the chance to get one over on their former employer. Tideland introduced a solar-powered buoy which, with the rise of off-shore oil platforms, proved so successful that Automatic was almost run out of business. Luddites take heed.

Demand for petroleum surged during the 1970s, which together with a restriction on the supply by the OPEC countries, drove prices sky-high. Not for the last time, oil companies found themselves awash with cash and looking for ways to invest their new-found liquidity. Solar power had obvious attractions. The oil industry is acutely aware that the petroleum is going to run dry one day, and if solar is truly to be the energy of the future, then it's natural that they would want to continue to enjoy their monopoly over its supply. During the next few years all the major oil companies started solar operations, and up until the 1990s Exxon, BP, Mobil and Shell were the biggest producers of solar panels.

Today solar energy can produce electricity for a wholesale cost of well under $3 per watt. The costs have reduced so much that the unit price of the solar panels themselves is less than that of the ancillary equipment involved in their installation. Other technologies have tried to enter the market – an alternative method using a glass laminate containing a thin film cell was actually the most popular system in the world in 2009 – but as prices for silicon have continued to reduce, it invariably reasserts itself quickly as the number one technology.

Over the years, there have been a number of developments that have enabled us to extract energy via the photovoltaic effect with much more efficiency. Individual solar cells produce a very low voltage, so several cells are wired in series inside a weatherproof container to make a photovoltaic module, or what we call a solar panel. The power that one module produces – a maximum of 400 watts in some cases, but often less than 100 watts – is seldom enough to meet the power needs of domestic or commercial users, so individual panels are usually grouped together to form a solar array. The energy generated by an array can be used directly at source or fed into the grid for use by others, or even stored in a battery. In the UK, most domestic generators feed some energy into the national grid, but the electricity they produce is in the form of direct current, which must be converted into alternating current using an inverter.[3] All these stages have an impact on the efficiency of energy conversion.

Solar panels produce most electricity when they are directly facing the sun, and their location on the Earth also affects their productivity. During cloudless days on the equator, where the Earth's surface presents a plane that is perpendicular to the sun, the rays beat down at a maximum strength of 2,400 kW/m² of solar radiation, or insolation, per day. In more temperate climes the sun shines with much less ferocity – between 700 and 1,000 kW/m².

Location makes a big difference. Industrialised production of solar power is growing rapidly. A solar farm is essentially a photovoltaic power station that can generate electricity on a utility scale. Completed in 2008 and covering an area of 110 ha, Waldpolenz Solar Park just outside Leipzig in Germany was the world's largest photovoltaic power system when it was constructed. It generated 52,000 MWh of electricity in 2011. The solar park in Topaz, south California is due for completion in 2015. It covers 260 ha and will generate 1,100 GWh of renewable energy annually. This variation is due to climate and location. In Germany each megawatt of capacity produces approximately 0.6 GWh in a year, while in California output is more than double at 1.4 GWh per megawatt, per year.

At latitudes away from the equator, PV arrays can be engineered so that they track the sun through the sky each day and tilt at different angles from the horizontal during the various seasons, thereby ensuring they always collect the maximum amount of energy. Currently these optimising systems for tracking and tilting are viewed as optional extras, but they can increase the amount of solar energy captured

---

[3] We think Thomas Edison would have found this highly amusing. Once he'd stopped laughing, he would then no doubt have tried to build a monopoly for himself in the inverter market.

by as much as 45 per cent. Most panels are fixed and their installation is set to provide the optimal output during the annual period of peak electricity demand. This is why in the UK and northern Europe you will find that panels tend to be situated only on south-facing roofs.[4] Almost all solar arrays producing more than 1 megawatt use tilting and tracking systems. Tilting can also increase efficiency over the long term. Solar trees, for example, as the name implies, are artificial solar arrays that mimic the look of their real-life counterparts. Solar trees provide shade during the day and at night can even function as street-lights. Famers can use them to provide DC power directly to equipment, and for communities in remote areas they could become the sole source of electricity, by providing a charge to storage batteries. Solar trees can produce 50 per cent more power than a flat solar power layout during winter, and 20 per cent more during other seasons.

The cells themselves are also extremely sensitive to shading. It takes only a tiny portion of the cell to be shaded from direct sunlight for the output to drop dramatically. This is due to the fact that shaded areas act as a short circuit, with electrons reversing into them rather than flowing through them. Instead of adding power to the cell, the shaded area absorbs power, converting it into heat. Sunlight can also be absorbed by dust, snow, or other obscuring material at the surface of the module. As this detritus accumulates over time it can have a surprisingly big impact on the performance of a cell. Google found that cleaning the flat solar panels on the roof of its corporate HQ every fifteen months resulted in a 100 per cent increase in output. However, simply by tilting the cells at an angle, the problem disappeared completely, as the panels were cleaned adequately by rainwater.

Taking all these issues into account, in Europe and those areas of the US on the same latitudes, solar panels have an average efficiency of 15 per cent, with the best models

---

[4] In the southern hemisphere they face north.

commercially available operating at just under 22 per cent. Given an average daily amount of five hours usable insolation, it's reasonable to expect a typical solar array located here to produce between 0.75 and 1 kWh/m^2 per day. By way of comparison, the same system located in the Sahara desert, with less cloud cover and a more favourable solar angle, would obtain around 8.3 kWh/m^2 per day (provided you could keep the sand off the panels). The Sahara covers more than 9 million square kilometres. It's a barren wilderness with precious little indigenous fauna and flora, inhospitable to human life and of little practical use. Covering just 1 per cent of the Sahara in solar panels – an area of 90,600 km^2 – would provide as much electricity as all of the world's power stations in 2012 combined.

The potential upside with solar is enormous, but before we get carried away with schemes to convert the Sahara into a global power station, there are a few major issues to consider. First of all, there is the question of getting the power to where it's needed, i.e. not the middle of the Sahara. To do that we'd need to construct a grid, and to transport the electricity to where it's wanted, thousands of miles away, would require cables thicker than our bodies. Even so, the losses in power over such massive distances would be immense, rendering the entire exercise completely impractical. Solar PV needs to be localised.

A further complication is that there's a theoretical limit to how much power we can produce from a solar cell. This maximum was calculated in 1961 by a physicist working in the semiconductor industry called William Shockley. The so-called Shockley–Queisser limit places the utmost solar efficiency at only 33.7 per cent. Current technology is approaching this limit. Although the best modern monocrystalline units on the market might convert sunlight into electricity only at around 22 per cent efficiency, much of this shortfall is due to the practicalities of design: light reflecting off the glass casing or shading caused by the thin wires on the surface of the silicon. More could be done to help existing single-junction cell technology to get as close to the limit as possible, but all the solutions are expensive, involving the collection of as much light as possible using the aforementioned tilting and tracking, or even a system of mirrors and lenses to concentrate the sunlight to an intensity of a thousand suns.

There are applications in which the Shockley–Queisser limit can be effectively overcome by using a different technology that employs multiple layers of semiconductor. In this system cells are stacked in series so that the radiation passing through the first layer may be absorbed by the subsequent layer (or layers). This makes it possible to capture light across the whole solar spectrum and is much more effective. If you could construct a cell with an infinite number of layers, you could collect 86 per cent of the light that hits it. In practice, the amount captured is somewhat short of this figure – and these cells are much too expensive to produce. To make them stack up commercially also requires the construction of a massive array of mirrors to concentrate the light.

There is still plenty more we can do with single-layer cells. The goal is to find a way of making ultra-cheap, single-junction cells that brings the price down by a factor of 25 to 100. This would make solar extremely competitive in comparison to the other sources of primary power, and help facilitate the market acceptance and dominance that we need. The challenge is that if we use cheap materials, i.e. those with smaller crystal grains, then the boundaries between the grains promote a process called 'recombination'. When an energetic photon has been absorbed by a semiconductor, and an electron/electron-hole pair is created, sometimes the pair inadvertently recombine before they can be drawn off to an external circuit. The more of these 'recombination sites' there are, the less efficient the cell is at producing electricity.

While there are many different and competing solar cells made from conventional semiconductors, there are also some promising technologies that, although they lose out in terms of absolute power and efficiency at the moment,

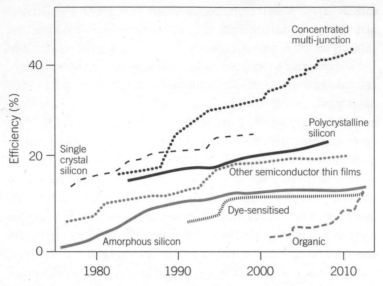

Fig. 7: A comparison of the best research-cell efficiencies for different PV technologies. (Source: US Department of Energy, National Renewable Energy Laboratory)

could lead to solar power production under much less stringent conditions in the near future. The common feature is that they are made of thin films, often on flexible substrates, comprising a sandwich of metallic electrode, semiconductor (organic, inorganic or a hybrid) and an uppermost transparent electrode.

Since 2010, the demand for thin film solar panels has fallen by almost a third, as sales of more efficient crystalline silicon versions (with their distinctive dark and light blue stripes) have displaced them. Thin film panels have enjoyed the lion's share of investment for the best part of four decades, due to the success of the first solar panels, but have failed to deliver the cheap solar power that has been promised for just as long, and are largely responsible for the fact that solar is still regarded as an alternative rather than a mainstream source of energy. Crystalline solar panels don't need to be sandwiched between two panes of glass, reducing weight and cost, while their solar-to-electricity conversion rate is 2 to 3 per cent better than thin film. This might not spell the end for thin film altogether. Gallium arsenide (GaAs) multi-junction cells, originally developed exclusively for use in outer space, are highly efficient and their use in terrestrial concentrators could provide the lowest cost in terms of dollars per watt and per kilowatt hour. These cells consist of multiple thin films produced using a highly complex process called metal-organic vapour phase epitaxy, which grows crystalline layers to create complex semiconductor multi-layer structures. They absorb nearly all of the solar spectrum, thereby generating as much electricity as possible. These are the most efficient solar cells produced to date, reaching a record high in April 2011 of 43.5 per cent. This technology is currently being used to power the Mars Rover.

As a whole, the market for all PV solar modules, whatever the underlying technology, is experiencing incredible growth due to the familiar issues of increasing energy demand, security of supply, rising fossil fuel prices and concerns over global warming. The global market grew in 2010 by 139 per cent, or 21 GW of increased solar power capacity. This rate of growth will not be sustained in the short term, but once grid parity is achieved (i.e. the point at which solar generates electricity at a cost less than or equal to the price of purchasing power from the main electricity grid), demand will increase rapidly once more. The opportunity for the UK is to ensure that it retains its world-leading research credentials in this field as demand ramps up. The most effective results will come from academics and industrial researchers working in concert across a range of fields and disciplines. To that end, a 'Supersolar Hub' of universities has been established to coordinate research activities. An interesting example is the initiative being led by CREST, based at the University of Loughborough and supported by the Universities of Bath, Liverpool, Oxford, Sheffield and Southampton. The consortium contains a deliberate balance of expertise, with no bias towards any one technology, and is active across all of the PV technologies including new materials, thin film chalcopyrite, crystalline-Si, thin film amorphous-Si, dye-sensitised solar cells, organic PV, concentrator PV, PV systems performance and testing.

※

There is more than enough sunlight to allow us to generate all of our electricity, but at present only a tiny fraction of the world's total electrical output (0.02 per cent in 2008) uses solar radiation as a fuel. While crystalline silicon is increasing in popularity, the pace of roll-out is being pegged back by the relatively long payback time, due to the high costs of production. However, there is growing interest in the development of PV based on organic materials (OPV) that can, in principle at least, be produced at extremely low cost over vast areas, using an in-solution-based process, requiring only a small energy input.

During the 1990s, there was great excitement about a new kind of organic material for electronics, which culminated in the award of the 2000 Nobel Prize for Chemistry to Alan Heeger, Alan G. MacDiarmid and Hideki Shirakawa. Their work paved the way for the development of organic light-emitting diodes (OLEDs) and organic thin film transistors (TFTs), which in turn led to the production of a brand-new range of flexible electronic devices, created by printing directly onto conducting plastic films. The organic materials used in most cases are polymers. Philips has already brought OLED technologies to market with display devices on consumer electronics while LG, Samsung, Sony and others have introduced OLED TVs and computer displays. Anyone reading this book on an e-book device like a Kindle also benefits from organic electronics, as the TFTs that drive these passive displays are also printed from organic materials. Given that a PV device is effectively an LED working backwards, it's hardly surprising that very soon OPVs were being developed. This is a highly active field of research across the globe and a number of key technologies are emerging, often from the same groups that originally worked on OLEDs, but also from the left field as well.

The first OPV had a single layer of organic semiconductor sandwiched between the pair of electrodes, and wasn't very efficient. The second generation had a pair of organic semiconductors, an electron donor and an electron acceptor, in a bilayer. These were an improvement, but as the current was generated from the interface, they were limited in their power by the size of the bilayer. When a photon excites the polymer and liberates an electron to leave a hole behind, the exciton produced can move only about 10 nanometres (nm) before the hole and electron recombine to liberate heat. To separate them and generate electricity there must be an interface between the two plastic donors and acceptors within 10 nm. The best morphology for this is a set of interpenetrating hairs (10,000 times thinner than human hairs). The construction leading the way at the moment is called a bulk hetero-junction (which looks very much like a mess of fine grains). Competition is important if this new technology is to gain traction in the market. Fortunately, there are many competing research groups around the world that have led to, or are leading to, spin-out private companies dedicated to the commercial exploitation of their findings. Alan Heeger, the Nobel Prize-winner in 2000, is involved with Konarka, a company that takes its name from a temple honouring the Hindu sun-god Surya. Konarka produces flexible solar cells based on bulk hetero-junctions with a certified efficiency of 9 per cent.

OPV is a young and developing sector that relies on a steady stream of talent migrating from university labs into industry. In the UK, much of the research in this field has its roots in the pioneering discoveries of Professor Sir Richard Friend and his team at the Cavendish Laboratory at the University of Cambridge. Friend's early work into organic LEDs and thin film transistors led to the establishment of several companies that are developing and exploiting the technology, such as Cambridge Display Technology and Plastic Logic. This Cambridge group is at the forefront of bulk hetero-junction OPV and is working with organisations around the world to drive the technology forwards. Eight19, which takes its name from the time it takes sunlight to reach the Earth – eight minutes and 19 seconds – is a developer and manufacturer of third-generation solar cells based on printed plastic. Originating from technology initially developed in Cambridge, these flexible, robust, lightweight solar modules benefit from high-speed manufacturing and low fabrication costs. With a fraction of the embedded energy of conventional solar modules, printed plastic solar modules are particularly well suited to consumer and off-grid applications and have found a market for pay-as-you-go electricity and mobile phone-charging in developing economies. Imperial College London has a thriving research group ploughing a similar furrow. Donal Bradley (a PhD student of Friend's) leads a large research team comprising chemists, physicists and engineers, and is on the board of the spin-out company Solar Press. Their bulk hetero-junction devices are based on organic materials printed on conducting electrodes.

As part of Project Sunshine, the University of Sheffield has a research programme into organic photovoltaics that was also begun by Bradley. This research is based on thin film blends of conjugated polymers and fullerenes, a further application of work carried out by Sheffield professor and Nobel Prize-winner, Sir Harry Kroto. Fullerenes are a new form of carbon. When light is absorbed by the semiconducting polymer, an exciton is formed that can be separated into an electron/hole pair by transferring an electron to the fullerene. By applying an electric field across the active film, the electrons and holes can be extracted to generate a photocurrent. In the same programme, David Lidzey's laboratory in the Department of Physics and Astronomy is exploring a range of new electron donor polymers and conjugated polymers (synthesised in Ahmed Iraqi's laboratory in the Department of Chemistry), along with fullerene acceptors for application in photovoltaic devices. The goal is both to understand the mechanisms that affect device efficiency and also develop manufacturing techniques to fabricate efficient devices. This is an example of scientists from different disciplines – in this case, polymer chemists, device physicists and chemical and process engineers – working together to create innovative technology. The resulting devices are tested in the Sheffield Solar Farm, on the roof of the physics labs, comprising a 58m² state-of-the-art silicon photovoltaic installation designed to measure the use of 'real-world' devices in northerly locations. The facility makes field-testing of a range of new and alternative photovoltaic technologies possible, some of which were made in the labs below from molecules synthesised across the street.

※

There is another set of technologies based on dye-sensitised solar cells. Here, a metal-organic dye captures the light rays and generates an electron, with the charge-separation process taking place across the interface with a titanium dioxide (TiO_2) semiconductor. These were first developed by Michael Grätzel at the Swiss Federal Institute of Technology in Lausanne. The team's major breakthrough was to use very finely divided TiO_2 in the form of nanoparticles, which have a very large surface area per unit mass (up to 300m² per gram of material). In effect, this gives 20g of TiO_2 a surface area bigger than a football field, which makes for highly efficient devices. The whole cell comprises a liquid or solid electrolyte over a layer of dye-coated TiO_2 that is structured like a carpet of tiny fibres. They can be made to be very thin and flexible through techniques as simple as screen printing. In 2009, a company called G24i began shipping Grätzel cells based on nano-TiO_2 and ruthenium-centred dyes.

Quantum dot solar cells are also based on the Grätzel architecture, but they employ low band gap semiconductor crystals that are so small they form quantum dots instead of organic dyes as light absorbers. These quantum dots have a number of unique features. Sunlight has photons of a wide range of energies. Only a certain amount of energy is required to liberate an electron/hole pair. This is the 'band gap energy' of a material. Some photons won't have enough energy to liberate an electron/hole pair and will simply pass through the cell as if it were transparent; others have too much energy. If a photon has more energy than the required amount, then the extra energy is lost. These two effects account for the loss of up to 70 per cent of the radiation energy. The minute size of the quantum dots means that the band gap can be tuned, or widened, simply by changing the particle size.

Another way to create more energy is simply to remove a layer of semiconductor. This stripped-down version of a normal PV cell has been found to boost energy output, and it promises a new route to cheap solar power. These cells are very inexpensive to make, and convert just under 11 per cent of light to electricity. These new cells are a twist on a technology called extremely thin absorbers (ETAs), in which the eponymous absorber is just nanometres thick. Liberated electrons travel to their electrode via a scaffold of semi-porous semiconductor TiO_2, while the holes travel through a conductor. A group led by Henry Snaith at the University of Oxford is developing new types of ETA cell that do away with the TiO_2 scaffold. The researchers believe that the electrons in this material are led on a meandering path, which drains too much of their energy. Instead they have developed a crystalline absorber called peroskite, through which they claim electrons travel faster and straighter, retaining more of their energy. By refining the scaffold further and using absorbers that liberate more electrons and holes from photons, the team reckons it can improve efficiency to around 20 per cent. At present, the cells cost between $0.70 and $1 per watt of capacity to make; but by scaling up production, the team believes the price could fall to $0.50, which would make it one of the cheapest emerging technologies on the market.

OPVs offer the potential for a truly mass-market solution, applied by nothing more sophisticated than a spray. However, there is still a lot of work to be done before we reach commercialisation. At present, the very best OPVs are less than half as efficient as standard crystalline silicon, with a much shorter operational life expectancy. There is as yet no consensus on how these issues can be addressed, with a variety of routes being explored. The central problem is that the materials with the highest efficiency in OPV devices are the ones with the lowest solubility, properties that limit their application via high-speed manufacturing processes – and as we've seen, many people are working on that. Some of the most promising research, being carried out by Project Sunshine at the University of Sheffield, involves engineering the chemical structure of low energy-gap donors to improve solubility and reduce production costs. The results should be a set of materials and a scaleable process for high-speed OPV manufacture.

☀

As a means of generating electricity, solar energy is, theoretically, nearly perfect. But the drawbacks that prevent it from being a truly holistic fuel are considerable. And even if we overcome these problems, scaling up solar power using photovoltaics will be hugely challenging. While the actual land area required is easy to calculate, we wouldn't want to do this all in one place. Sites would be distributed widely, and the size of this undertaking shouldn't be underestimated. Even if solar cells could be produced cost-free, that still wouldn't solve the problem of producing the amount of carbon-free energy we need by 2050. Assuming a mean efficiency of 10 per cent conversion, which is entirely reasonable outside equatorial latitudes, putting a solar energy convertor onto the roof of every home in the US would generate just 0.25 TW of power. Covering every roof in the world with solar panels will still not get us anywhere near the 10 to 20 TW we need. To get that kind of power will require more than one approach, and not just to solar but to the production of energy as a whole.

Moreover, photovoltaics produce electricity – a vector

– and we have seen that there's no viable, cost-effective way to store it. So we must also consider what we're going to do with all this energy we're producing inexpensively from sunshine. We could feed it all into the main supply, but if we do, we encounter further problems. Most solar arrays are connected to the main grid, but not all. It's these independent systems that best illustrate the problems with solar power: intermittency (as with all forms of renewable energy) and insolation. Standalone systems vary widely in size, from the units that power a watch or calculator to massive arrays that power isolated buildings, communities or even spacecraft. Sunlight is intermittent, but our energy needs are continuous. The sun doesn't shine 24 hours a day, nor does it shine much during the long winter months when our energy needs are greatest. Solar power is plagued by inefficiencies, but unfortunately it's also the only way for us to go.

The most profound challenge in meeting rising energy demand, in the face of climate change, is that of peak oil. The development of small-scale technologies for energy conversion and energy efficiency is an essential component of the combined strategies to confront the challenge of energy generation and fuel supply as a whole. Technical progress in the field of solar is currently swift, with each new development promising leaps in cost reduction, efficiency or flexibility of application. However, these technologies will have an impact in the real world only if they are successfully integrated into the lives and environment of ordinary people.

First-generation PV is well established as part of the low-carbon strategy in several countries, most notably the highly advanced economies of Japan and Germany, where its application is extending rapidly as a result of government support and a mix of public and private investment. In spite of this, PV's potential – even at today's technological standards – is vastly under-realised and under-used. Solar is a far more efficient means of generating electricity than wind, water or biomass; it's far more versatile and its suitable application much more widespread: the sun does, after all, shine everywhere at least some of the time. Yet realising solar's potential will take a significant reduction in costs, together with a massive increase in manufacturing output. OPV is one emerging technology that promises to deliver on both counts; the other is the luminescent solar concentrator (LSC), in which manufacturing methods employing low-cost raw materials and roll-to-roll or high-speed sheet deposition are the focus of significant effort. A study into the effectiveness of this technology is taking place in social housing projects operated by Sheffield City Council and urban high-rise buildings in Bangladesh. These locations present users with not only cultural differences but differences of energy infrastructure and norms of energy use, and radical differences in built environment and tenure. The results are critical in determining how step reductions in cost for next-generation technologies have to be balanced against a reduction in intrinsic stability of organic materials when compared to crystalline silicon.

Many people think of solar power as simply a means of generating electricity, but it's much more than that. One day it will be the source of all the energy we consume. But before we move to an exclusively solar economy, we need to address the issues of intermittency, transmission and portability. In short, finding efficient and cost-effective ways of storing and transporting sunshine is the final piece of the jigsaw. And we need to find them quickly, because, both literally and figuratively, solar power needs to happen overnight.

13 Whatever Gets You Through the Night

> 'I've been told that I have a lot of energy. The secret
> is that I use renewable resources. Some days I'm
> solar-powered. Some days I'm wind-powered. And
> some people in this room might think I'm hybrid gas-
> powered. You'll just have to guess which it is today.'
> *Bill Richardson, US Energy Secretary, 1998–2001*

It's difficult to comprehend the enormous numbers involved
when we talk about energy. According to the CIA *World
Factbook*, in 2004 electrical energy consumption from all
sources – solar, nuclear, renewables and fossil fuels – was
15,406 TWh, but we can also express this figure in terms of
the amount of oil that contains the same amount of energy. In
this case, that would be 1,205 million tonnes of oil equivalent
(Mtoe). And remember this is just the demand for producing
electricity. If we add in heating, transport and agriculture,
the figure for the same year rises to 9,336 Mtoe. In many
ways it's more helpful to think of energy consumption at a
micro scale. In most European nations the average electricity
consumption is about 2,000 KWh per person per year, but
it ranges across the continent from 7,467 KWh per person
in Norway, where electricity is inexpensive and people use it
for heating their homes, down to just 352 KWh per person
in Romania, where household incomes are much lower, electricity is relatively expensive and biomass or coal is used for
heating. In the US, average consumption is 4,387 kWh per
person per year. For these figures it's possible to work out
that a 1,000-MW power plant working at 80 per cent capacity will provide enough energy for 640,000 homes in the
USA or 1.3 million European households.

This demand is not constant, but varies at different times
of day. Figure 8 shows the demand profile of a typical UK

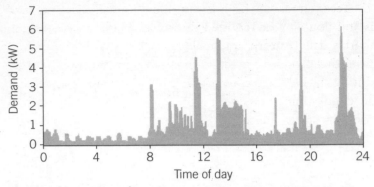

Fig. 8: UK individual household electricity demand. (Source: The Environmental Change Institute, University of Oxford (modified))

household, monitored every two minutes. Matching such a spiky demand profile to generating capacity would be extremely difficult, but fortunately the aggregate demand when we add in industrial and commercial consumers tends to be much smoother and follows patterns that are generally predictable, as in Figure 9.

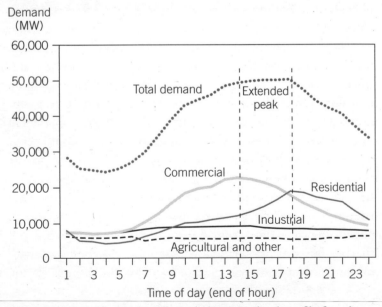

Fig. 9: Example of an aggregated electricity load profile for a hot day in California. (Source: Lawrence Berkeley National Laboratory)

Yet even taking this into consideration, the daytime load is still double that required during night-time. In addition to this daily pattern are smaller, long-term, seasonal variations that produce a greater demand for heating and lighting during the cold, dark winter months or, in some regions, for air-conditioning units during the hottest summer months. In some instances, industrial users may follow an annual cycle that is separate to these seasonal variations, and this further affects the aggregate supply. Electricity suppliers rely on a combination of experience, local knowledge and hard data backed up by information about economic performance and growth to make accurate predictions about demand. On top of this, the utility is expected to allow for 'acts of God', natural disasters and freak weather conditions to ensure continuity of supply.

Using fossil fuels, which can be stockpiled, transported and easily ramped up and down, matching load to demand across a grid is challenging – not least because the timing of peak demand rarely coincides with the availability of the most convenient peak supply – but possible. Even if we could generate enough power, attempting to do this using the various renewable forms of energy exclusively would be nigh-on impossible. Renewables are essentially intermittent sources of energy. Sometimes the cause of intermittency can be quite predictable: for example, the sun doesn't shine at night, while tidal power is reliant upon, well, the tides. To use an intermittent energy source effectively requires displacing non-renewable fuel that would otherwise be used – by topping up the power to the grid, for example, or by storing the captured renewable energy for use when needed. The storage of energy to make up the shortfall caused by intermittency is vital if we are to guarantee a continuous supply. The big question is: how do we store all the energy we are capturing?

※

If electricity from solar is to be supplied outside the hours of insolation, then it needs to be stored and, at the moment, that means a battery. We have already discussed the shortcomings of battery power, but in the case of non-portable applications up to the size of a building, lead acid batteries can be used with a reasonable degree of effectiveness. Lead acid batteries are cheap and robust enough to withstand the constant charging/discharging cycle. However, lead batteries are inefficient and certainly no solution for energy storage on a local, let alone a regional or national scale. This is a significant drawback, but it is surmountable.

At the moment almost all the energy we store is in the form of chemical bonds. While we are adept at converting kinetic energy (wind and wave), electromagnetic radiation and heat into electricity, we have not yet managed to convert any of these forms of energy into stored potential energy at scale. In fact we get most of our energy by breaking chemical bonds apart in fuels such as oil and coal, which are by far the cheapest sources we have. In theory, if we could produce enough biomass, the best way to extract the energy from it would be to convert it into carbon-neutral ethanol or biodiesel and burn that instead. If we could manage solar energy in the same way as chemical bonds we could store the power until we needed it. All that's missing is a means to capture, convert and store the electricity that solar produces.

Gemasolar, located just outside Seville in Spain, is the world's first solar power plant to provide energy 24/7 on a commercial scale. At the heart of the 185-hectare site stands a tower 115 metres high, surrounded by 2,650 adjustable mirrors called heliostats. Each heliostat has a surface area of 120m^2 (1,292 sq. ft.) that focuses the sunshine on a solar receiver and steam turbine at the top of the tower. The solar receiver produces superheated steam at 275°C, which drives the turbine to produce electricity via a generator. The energy conversion rate of solar to electricity is 17 per cent. Unlike conventional solar power stations, Gemasolar can produce electricity through the night as well. A large proportion of the solar energy it receives is used to heat up molten salts – a mixture of potassium and sodium nitrates – to a temperature of 500°C. This heat can be converted into electricity for up to fifteen hours without sunlight. In June 2011 Torresol Electricity announced that the plant had achieved a full 24 hours of solar energy production. It has a 120-MW capacity, and its goal is to provide uninterrupted electricity throughout the summer. The electricity it produces is expensive in comparison to plants using fossil fuels, but as the technology develops, costs are predicted to fall to between 6 and 10 cents per kilowatt hour.

The system at Gemasolar is a novel and effective example of how thermal storage can make solar energy manageable. It's an alternative to the electrical battery storage systems that have long been employed to stabilise power across distribution networks. In Puerto Rico, for example, electric power is managed across the whole island using a battery capacity of 5 MWh; and in Fairbanks, Alaska a 6.75 MWh nickel-cadmium battery bank has been used to stabilise power since 2003. But batteries are expensive to produce and maintain, and, due to the crystallisation process that results from the charge/discharge cycle, have very limited life-cycles. The pure crystals that form cannot be re-dissolved back into the electrolyte, and can grow large enough to adversely affect the interior structure of the battery itself. Gemasolar uses a different process of energy conversion: storing energy as heat converted from solar radiation, rather than as chemical energy from electricity.

There are other possible technologies for large-scale battery storage. Donald Sadoway is Materials Chemistry Professor at the Massachusetts Institute of technology, and his team has focused on ways of producing more efficient batteries. He is the co-inventor of a solid polymer electrolyte, which could be used to produce batteries with twice as much power per kilogram as is possible in the best lithium ion models in today's market. These 'flow batteries' have a number of advantages over traditional models. Power and energy components are separated, so there are no solid-state phase transitions. The result is a flexible layout, a long life-cycle, rapid response times, no harmful emissions and no need to ensure that all cells have an equal charge by over-charging. Maintenance is low and flow batteries have a very high tolerance to over-charging and over-discharging. But more important, perhaps, are the revolutionary molten salt batteries that are now being developed based on his research. The magnesium and antimony electrodes in the battery are separated by a molten salt, and could provide a very low-cost stationary energy storage system. Experimental data shows a 69 per cent DC-to-DC storage efficiency, with high storage capacity and very low leakage.

Inevitably it's not all good news. Flow batteries are very complicated in comparison with standard batteries, requiring pumps, sensors, control units and secondary containment vessels, while their energy densities, which vary consider-ably, are much lower than comparable lithium ion products. Redox flow batteries, which employ a reduction-oxidation process to charge and discharge, are being tested on wind farms in Huxley Hill, Australia and Sorne Hill in Ireland.

Another highly promising battery technology is the metal-air battery. Lithium-air (Li-air) batteries use a chemi-cal process that involves the oxidation of lithium at the anode and the reduction of oxygen at the cathode to create a flow-ing current. The basic science has been understood since the 1970s, when Li-air batteries were proposed as a power source for milk-floats and other electric vehicles, but advances in the quality of available materials together with the consider-able energy and environmental benefits have rekindled inter-est in the last few years. The heart of the Li-air battery's attraction is its extremely high energy density, due to the fact that oxygen is drawn in externally from the air instead of storing it in an oxidiser, which gives it an energy per vol-ume equivalent to that of a petrol engine. The technology is still in its infancy, with four competing Li-air prototypes being actively researched for application in mobile phones, laptops and hybrid cars. The main issue to be overcome is that currently, side reactions with the electrolyte cause degradation and lead to poor performance through repeated charging and discharging cycles. Moreover, while Li-air batteries work by exposing a lithium anode to an electrolyte that grabs positively-charged lithium ions and pushes them toward the cathode, made of a porous material that allows oxygen from the air to form the crucial lithium peroxide, there isn't a good way to reverse the process. However, in June 2012 a team led by Peter Bruce, a chemistry professor at the University of St Andrews in Scotland, announced that by using an ionic liquid for the electrolyte they had solved the instability problem, while a nanoporous gold electrode was giving the battery faster operating characteristics compared with traditional porous carbon electrodes. The solution is not sufficiently cost-effective to be commercialised, but it does prove that a lithium-oxygen battery capable of sustained cycling is possible.

It's likely that both Li-air and Mg-Sb batteries will play a big role in the near future. Mg-Sb is certainly the front-runner for car batteries (magnesium carries two electrons to lithium's one, which should make it more efficient) and other portable technology, but once the problems are overcome, Li-air is likely to be used for large stationary energy storage and other non-portable activities, where its high energy density offers a real advantage. But this optimism must be tempered. Despite these promising developments, batteries are unlikely ever to provide an effective energy-storage solution for anything bigger than a single building. They are simply too inefficient, whatever the chemistry, ever to provide a viable, cost effective solution to intermittence at even a local level, let alone the regional or national scale we require. While the mobile phone and AA battery can store around 700 J/g and are much better storage media than the lead-acid car battery, they all pale in comparison to a cheese and bacon sandwich or a chocolate chip cookie, which, at 2,100 J/g, are three times more energy-dense. It's not that these unhealthy snacks are a highly effective energy store – rather that the batteries are just so poor.

☀

Storing energy as electrochemical potential is not very mass-efficient, but storing it as oxidisable chemical bonds is. What we need instead of a battery is a liquid fuel. As we've seen, all that's missing is a means to capture, convert and store the electricity that solar produces. We could use the electricity to run electrolysis, splitting water into its components to make hydrogen and oxygen. Another option is to make hydrogen from water directly using a two-axis parabolic dish. On the plus side, these dishes can be 30 per cent efficient; but they require catalysts made out of the scarce and expensive element platinum. Yet that's not the biggest problem preventing this technology from scaling up to our 10–20 TW needs. To reach even the bottom end of our target, we'd have to build one of these dishes every second for the next 50 years, using the world's entire supply of steel and aluminium in the process – and we'd have run out of platinum long before then anyway.

In any case, hydrogen is not an ideal vector and unless it can be used at the point of production, without the need for transportation, it's questionable whether we can put it to any practical use at all. The problem with hydrogen is that it's a gas at atmospheric pressure. It can be stored as a gas, but it has a very low density and needs a lot of space; or it can be compressed and stored as a liquid, but that needs either very high pressures and massive equipment or very low temperatures and massive equipment. Or it can be absorbed into special molecular cage structures to make hydrides and then desorbed by heating when it's required – and in every case the total hydrogen storage system has a lower energy density than a liquid fuel.

However, by combining the H_2 we collect with CO_2 we can produce methanol, which is a liquid and much easier to contain, transport and distribute. The reactions are quite simple: three hydrogen molecules and one carbon dioxide molecule combine to make one methanol molecule and one water molecule ($3H_2 + CO_2 \rightarrow H_2O + CH_3OH$). Methanol itself can be converted into other fuels or used itself directly as a liquid fuel. Because the hydrogen comes from a renewable source and atmospheric carbon dioxide is fixed to generate the fuel, the methanol produced would be effectively a carbon-neutral fuel. To scale up this use of solar power will require the development of technology that can not only absorb and convert sunlight into electricity, but also cost-effectively produce chemically usable fuels in order to deliver this energy to end-users.

To achieve that we need to go a stage further. Photovoltaic cells are essentially artificial photosynthesis systems: they use light to separate charge. They operate on the same principle as plants, but use inorganic materials (silicon wafers and wires) instead of leaves and branches. If we could modify photovoltaic systems to produce fuels, couldn't we do the same with natural photosynthesis – namely, bio-engineer plants that will do the same thing? This is arguably the greatest challenge for chemistry in the 21st century, but if we can achieve it, then we really will have a clean alternative to oil on tap.

Photosynthesis, the most efficient way to turn sunshine into energy, has probably existed on Earth for around 3.4 billion years. Plants create vast amounts of energy by simply soaking up the sunshine. The big advantage plants have over our photovoltaic systems is that they can store the energy and use it whenever they like: trees don't suffer from intermittency because they make their own fuel and store it until they need it. If we could store solar energy in a similar way and make it transportable, building vast solar farms in the Sahara would begin to make sense. Inevitably, it's proving easier said than done. Photosynthesis took plants millions of years to evolve, but even they're not very good at it. It works by using the sun's energy to separate water into its constituent parts, hydrogen and oxygen, and then combining them with atmospheric carbon dioxide to produce the energy-storage molecules: sugars and starch.

In plants this whole process is lousily inefficient, so much so that they could hardly have provided us with a worse base-platform to engineer from. For a start, they are the wrong colour. Vegetation's green hue means that only a few per cent of the solar spectrum can be absorbed. Trees should really be black. This limitation also means that if the sun shines too intensely, the plant's system gets overwhelmed and production is shut down completely after just 30 minutes. What energy plants do capture is stored as solid carbohydrates, which with their relatively low energy density are hardly the best medium for our purposes. And so it transpires that bio-engineering photosynthesis isn't just about replicating the process but completely redesigning it.

In principle the science is quite straightforward. We need a version of a photovoltaic cell to absorb light energy to liberate electrons, called an 'antenna'. Then these electrons must be guided by catalysts towards the right molecules so that they can react to produce the kinds of fuel we need. In practice, guiding the electrons and producing the right fuel is a complicated business. Commercial viability requires a system that is efficient, durable and cost-effective, and we're some way off achieving that, but breakthroughs have been made. In 2011, the team at Sun Catalytix, led by Dan Nocera of the Massachusetts Institute of Technology, announced that it had created a wireless 'artificial leaf'. The leaf can make hydrogen from ordinary water using catalysts that are relatively cheap and efficient: a blend of nickel, molybdenum and zinc for hydrogen, and cobalt borate for the oxygen. The cost of the hydrogen is not yet commercially viable at $7 per kilogram (by comparison, making hydrogen by reforming methane costs just $2.50 per kilogram), but the aim is get the price down to less than $3 per kilo within a decade.

Nocera is just one of an army of solar scientists searching for the breakthroughs that really are going to transform the world we live in and the way that we produce and consume energy. At Imperial College London, a team of researchers led by James Barber has embarked on a £1 million project called Artificial Leaf, which aims to work out exactly how leaves use sunlight to make useful molecules and then to mimic this process artificially to create hydrogen and methanol. Similar projects are running across the world. The Dutch government has allocated €40 million for similar research, while in the USA $35 million from the Federal Reserve is being allocated for projects that can create fuels from sunshine. If any of the systems being researched can reach a conversion efficiency of just 10 per cent, then only 0.16 per cent of the world's surface would be required to provide the 20 TW of energy we require. And unlike their biological equivalents, these artificial leaves could be located in arid deserts, thereby negating the need to appropriate valuable agricultural land to produce biodiesel. Nocera has calculated that using artificial leaves to split a few litres of water a day into hydrogen and oxygen would be enough to meet the energy needs of every household.

Scientists can already produce hydrogen by splitting water into its elemental components, but this it achieves using materials and processes that are too expensive to allow commercialisation. A critical challenge is developing catalysts from cheap, widely available materials in ordinary conditions. In April 2012 one of the leading researchers in this field, Nate Lewis, Professor of Chemistry at the California Institute of Technology in Pasadena, summed up the situation: 'You've got this three-legged stool: a system has to be efficient, it has to be cheap, and it has to be robust. Pick any two criteria and I can give you that, but not all three.' Fortunately, there is significant funding going into finding a way to deliver all three. Lewis is also the head of the Joint Centre for Artificial Photosynthesis, supported by $122m from the US Department of Energy, and his is not the only team looking for an answer. We are still at the stage of 'promising developments' rather than 'amazing breakthroughs', but Lewis is confident: 'We're like the Wright Brothers. Our job is to fail fast and fail frequently and move ahead.'

John Loughhead, executive director of the UK Energy Research Centre, is equally positive about the possibilities for artificial photosynthesis: 'We know that plants have already evolved to do it and we know that, fundamentally, it's a workable process on a large scale. Ultimately, the only sustainable form of energy we've got is the sun. From a strategic viewpoint, you have to think this looks really interesting because we know we're starting from a base of feasibility.'

The real goal however, is to be able to mass-produce photosynthesising particles – perhaps nano-scale semiconductors – that can be simply immersed in water and placed in sunlight to produce fuel at ultra-low cost. James Durrant, a colleague of Barber's in the chemistry department at Imperial College, has recently developed a catalyst from rust that carried out part of the water-splitting reaction. The process is too inefficient at the moment but it does work, and Durrant's team believe they can improve the performance significantly by engineering the rust particles through adding small catalytic amounts of cobalt onto the surface. Sun Catalytix is also working on a catalyst. Made from cobalt and phosphorus, it can split water at room temperature.

This kind of efficient water-splitting technology is likely to prove the best way of storing solar energy. By day artificial leaves will use the energy from sunlight to split water into hydrogen and oxygen, and by night the hydrogen can be burned to generate electricity at the point of production, thereby avoiding the issues of storage and distribution. Hydrogen only really comes into its own as a fuel in these circumstances: when production is small-scale and the fuel is burned in the same location. The energy density of hydrogen (in J/g) means that it's many, many times more efficient as a storage medium than even the most advanced batteries. Working prototypes are due within a couple of years, but we shouldn't expect them to be cheap, and it could be fifteen years before one of them is market-ready. Yet we can be confident that they will appear.

Indeed, building on this research, and with just a little more complex chemistry, we could potentially be pulling carbon dioxide out of the atmosphere to produce liquid methanol and dimethyl ether, which is a much more complex fuel than hydrogen and also easier to transport.

✺

Liquid solar fuel can also be produced using photosynthesising micro-algae. All strains of algae produce a lipid – or fat – that has properties very similar to diesel. The lipid content varies, but strains with lower content may grow as much as 30 times faster than those that are lipid-rich. The challenge is finding a strain with a combination of good yield and growth rate that is not difficult to harvest. Algae can produce 300 times more biofuel than any of the conventional biodiesel staples: rapeseed, palms, soya or sugar cane. Using the fuel produced does release carbon dioxide into the atmosphere, but unlike fossil fuel that carbon dioxide was taken out of the atmosphere by the growing algae and so it can be considered carbon-neutral. Algal fuel has many attractive characteristics: its production requires virtually no fresh water, using sea-water or waste water instead; it can be grown on arid land, or land with an excessive level of salinity; it's also entirely biodegradable, and relatively harmless if spilled. Supporters of algal biofuel make some impressive claims for its potential. The US Department of Energy estimates that the domestic replacement of petroleum with algal biofuel would require 15,000 square miles: some 0.42 per cent of the nation's landmass, or around 14 per cent of the area currently used to grow corn. Algae grow 20 to 30 times faster than food crops and have a rapid harvesting cycle of just 1–10 days, which means many harvests are possible in a very short time-frame. Unfortunately, many of these claims remain unrealised today, although Mary Rosenthal, head of the Algal Biomass Organization, believes that a grant of production tax credits is all it would take to achieve full commercialisation and price parity with oil by 2018.

Algal biofuel has long been a dark horse in the race to find an alternative to petroleum, and there are still several hurdles to overcome before it can step out of the shadows and become a genuine contender. In October 2012, the US National Research Council (USNRC) issued a report questioning its sustainability. The burgeoning industry has been plagued by questions about the levels of resources it uses and whether it can ever be practically ramped up, but the USNRC report brought these into sharp focus by determining that even scaling up to just 5 per cent of the domestic transportation fuels market 'would place unsustainable demands on energy, water and nutrients'. Estimates regarding the level of those demands vary to a remarkable extent. For example, to produce a litre of algal biofuel requires between 3 litres of fresh water (which is comparable with that required to produce a litre of petrol) and 3,650 litres of fresh water. This huge difference arises from the multitude of production methods: for example, whether algae is grown in salt water, ponds or bioreactors; where the sites are located; rates of evaporation, and so on. The report was also concerned about estimates of the amount of fertiliser required to produce 39 billion litres of fuel: between 44 and 107 per cent of the country's total nitrate use in 2011; while phosphorus requirements would run at somewhere between 20 and 51 per cent of the total domestic phosphorus market.

None of these problems is a show-stopper for algae-based fuel. As the USNRC concluded:

[We do] not consider any one of these sustainability concerns a definitive barrier to sustainable development of algal biofuels because mitigation strategies for each of those concerns have been proposed and are being developed. However, all of the key sustainability concerns have to be addressed to some extent and in an integrative manner … if the promise of sustainable development of algal biofuels has any chance of being realised.

Producers do need to work smarter. Facilities need to be designed so that they minimise water-use and recycle nutrients and wherever possible tap into waste streams that can provide both. Science has a big role to play in helping algal biofuel step into the light, both in terms of the industrial processes and the cultivation of new algal strains themselves.

Algae were originally cultivated in large open ponds. The main problem with this system was that open ponds are prone to contamination, so only the hardiest strains of algae were used: those that could thrive in adverse conditions and out-compete invasive species. These were not necessarily the strains most effective at producing lipids. Open systems using a monoculture are also vulnerable to viral infection. This combination led to highly unattractive yields, and open-pond systems have been almost entirely given up for the cultivation of algae with high oil content. Today the most common system of algae cultivation involves a photo-bioreactor (PBR) into which nutrient-rich water is pumped through bioreactors (tubes of plastic or borosilicate glass) that are exposed to sunlight. This process is costly and difficult to operate, but it provides high levels of control and productivity. PBRs can operate in deserts or areas where the groundwater is saline, and can even be located off-shore. In essence, ponds are cheaper (in both financial and carbon terms) but suffer from contamination issues, meaning that only microalgae that can grow under extreme environmental conditions (such as high salinity or high pH) can be routinely cultured. In contrast, PBRs do not suffer from contamination problems and a much wider range of microalgae can be cultured under well-controlled environmental conditions to higher densities in outdoor PBRs using sunlight. However, PBRs are more expensive to run and more difficult to scale up.

The failings of the open-pond system and the specialist equipment required to operate a PBR at scale are the main reasons why widespread mass production has not been realised. A third system – called the closed-loop (CL) – seeks to address these issues. As the name suggests, CL systems avoid the problem of contamination by organisms blown in from the outside world because they are not exposed to the open air. Up until now, the main issue has been finding a cheap source of sterile carbon dioxide to feed the algae. There is evidence that carbon dioxide from industrial chimneys could be put to such a use. The University of Sheffield is involved in a series of pilot projects that it hopes will lead to the development of an integrated, flexible, multi-purpose system. Across a broad field of disciplines, scientists and social scientists are researching all aspects of algal fuel production, from costs and feasibility to public reaction and global impact. Microbiologists are identifying the best algal species for each geographical location and purpose; molecular geneticists are tailoring the physiology and metabolism for each need; organic chemists are identifying and monitoring novel products of algal metabolism; while engineers are developing the infrastructure of the algal farm. It may not be too long before we see huge coastal farms, drawing seawater for the culture of algae, and producing biofuel and high-value chemical feedstocks. At the same time, these installations will concentrate mineral fertiliser from seawater and remove pollutants, including carbon dioxide from nearby power plants. The whole activity could be powered sustainably by solar energy.

But the cultivation of algae is only half the problem. There remains the substantial issue of separating the useful lipid from the rest of the organism. There is currently no cost-effective method of harvesting and removing the water from the algae for it to be processed adequately. A possible solution to this was announced in January 2012. Will Zimmerman, who leads a team in the Department of Chemical and Biological Engineering at the University of Sheffield, has developed an inexpensive way of producing microbubbles that can float algae particles to the surface of the water, making harvesting easier and cheaper. A pilot plant is being built to test the system on an industrial scale at Tata Steel's factory in Scunthorpe, using carbon dioxide from its flue-gas stacks. Also involved in the project are Jim Gilmour from the Department of Molecular Biology and Biotechnology and David Lidzey from the Department of Physics and Astronomy, whose work focuses on another limiting factor to algae growth in CL systems: the provision of light. For the mass cultivation of algae, sunlight is essential – any type of artificial light would be much too expensive. However, the spectrum of sunlight contains large parts that are not absorbed by the pigments of algae and thus go to waste. Gilmour and Lidzey suggest using fluorescent dyes that absorb wavelengths of light currently under-used by algal cells and that emit longer wavelengths of light tailored to the absorption peaks of the light-harvesting pigments, thus enhancing the rate of photosynthesis. If successful, storing light energy emitted from the dyes during the day and re-emitting it at night could allow photosynthetic productivity to continue during the hours of darkness.

If algal biofuel remains a good outside bet as a genuine alternative to either petroleum or the proposed hydrogen and ethanol (biodiesel) economies, then the same can be said for methanol. The difference is that while the odds on algal fuel are shortening, methanol is a former red-hot favourite that has drifted down the field. A decade ago, all the smart money was on methanol, the fuel that propels racing cars around the Indianapolis 500, replacing petroleum. Methanol (CH_3OH) should not be confused with ethanol (C_2H_5OH). They are closely related compounds – both are chemically alcohols – but with different uses and properties. Methanol is poisonous if ingested and is generally derived through synthetic chemical processes, whereas on a commercial scale ethanol is created by fermentation of food crops. Ethanol is poisonous in large quantities but relatively harmless in small doses, and is the 'alcohol' found in beers, wines and spirits. Both substances can be used as stores of energy, but methanol's use as a motor fuel in the US has been largely phased out after a promising start, in favour of ethanol produced from starchy crops such as corn and sweet potatoes. Yet, for many years, methanol was considered a serious contender.

In 1980, the state of California began a pilot programme that allowed anyone to convert a petroleum vehicle to one that ran on 85 per cent methanol. More than 500 vehicles were converted to run on a high-compression mix of 85/15 methanol/ethanol. The results were impressive but the anticipated market for flexible-fuel vehicles (FFVs) failed to emerge, in part because gasoline prices remained low for the rest of the century. It was still something of a surprise when in 2005, with the market for oil shifting and gasoline prices soaring, the then Governor of California Arnold Schwarzenegger curtailed the use of methanol after 25 years and 200 million driving miles, in favour of expanding the market for ethanol, which was considerably more expensive to produce. Seven years later and the utility of ethanol is the subject of such fierce debate that its future looks doubtful. Five billion bushels of corn are used for ethanol production each year, a substantial proportion of total US production, and this is contributing to a global hike in food prices. Consequently, methanol, 'the other alcohol', is once again mounting a challenge to be a fuel of the future.

Given the convenience and high energy density of liquid fuels, the internal combustion engine is likely to be a cornerstone of personal transport for many years to come. One of the more interesting recent developments in this field comes from the British company Lotus. Their Omnivore is a two-stroke direct-injected engine designed to take advantage of the latest in electronic engine management and to run on just about any liquid fuel (hence the name). It uses an air-assisted direct injection system. A movable 'puck' in the top of the cylinder head allows the compression ratio to be varied to accommodate a range of fuels. There is also a feedback system that uses the composition of the exhaust gases to detect the nature of the fuel so that the fuel/air ratio and the compression ratio can be optimised. Depending on what fuel is used, the Omnivore can be spark-ignited (for gasoline, ethanol and methanol) or compression-ignited (for diesel, DME (dimethyl ether, a diesel substitute) or vegetable oil). It's also capable of compression ignition for fuels that normally require a spark. This gives diesel-engine-like performance and efficiency for light fuels without the particulate and nitric oxide emissions that require highly expensive after-treatment systems in standard diesel engines.

Together, the many advantages of liquid fuels provide the salient reason why, despite the predictions of numerous researchers, governments, companies and Jeremy Clarkson, our future energy economy is unlikely to be based on the universe's lightest element. The proposed hydrogen economy would use the gas to store renewable energy and power fuel cells, internal combustion engines and even electronic devices. Notwithstanding the impracticalities of actually producing the hydrogen itself, creating an infrastructure capable of handling the necessary transport and distribution has proved too costly and downright difficult to get up and running.

One prominent champion of methanol is 2006 Nobel Chemistry laureate George Olah. He proposes that modifying our existing petroleum infrastructure to run on methanol will prove easier and cheaper than trying to develop a remote distribution system that can cope with an explosive and almost uncontainable gas. The key breakthrough here has come in the production of methanol. The easiest way to create methanol is by steam-reforming methane, produced when syngas, a mixture of carbon monoxide and hydrogen, is turned into liquid hydrocarbons through a collection of well understood chemical reactions called the Fischer–Tropsch process. Methanol can also be produced directly from syngas, as the steam reformation process (used to refine fossil fuels into more useful hydrocarbon products such as diesel or gasoline) also produces carbon monoxide. While the chemistry is simple – one carbon monoxide molecule plus two hydrogen molecules give one methanol molecule ($CO + 2H_2 \rightarrow CH_3OH$) – the process is not. It's also energy-intensive. The thermodynamics and plant design are thoroughly mapped out and are already used widely in coal liquefaction and the production of biodiesel from biomass. Indeed, the fuel itself has a long practical history. The Germans synthesised methanol to provide rocket fuel for their V1 and V2 programme, while the South Africans did much the same to keep their vehicles running during the years of apartheid and international isolation. However, in both cases the methanol was produced from coal, so it could hardly be considered a green alternative to fossil fuel, rather a different storage medium.

The issue is that, like petroleum, electricity and hydrogen, methanol is a vector for energy storage and does not occur in large natural deposits. There are no reserves for us to tap into, so the objective for a renewable energy economy must be to ensure that the ultimate source of fuel, the syngas, comes from a carbon-neutral source. In 2009, Scott Barnett and colleagues at Northwestern University in Evanston, Illinois demonstrated that the process typically used to split water into hydrogen and oxygen, a solid electrolysis cell, could be the answer. Barnett's process involves combining equal parts of hydrogen and carbon dioxide with two parts water within the device, which at peak conditions can produce $7 cm^3$ of syngas per minute for every $1 cm^2$ of the electrolysis cell's surface. Turning this syngas into methanol is a straightforward industrial process. The method requires a constant supply of water vapour and carbon dioxide, but both gases are released when the methanol is burned and could be re-used. While the gas-capture system required adds additional cost, it would be cheaper than the equivalent technology for hydrogen, which uses fresh water that is heavy, and therefore expensive, to transport.

Critics point to the obvious drawback in this system: the energy carried by methanol produced in this way is less than was needed to create it. But this somewhat misses the point.

The real opportunity is to use methanol to store energy from renewable sources that are producing more electricity than is required. All those wind turbines and wave farms generating power when we don't need it, could be storing this energy as methanol: a vastly more efficient medium than any battery, chocolate cookie or cheese and bacon sandwich.

There are so many potential sources of syngas that methanol production doesn't need the appropriation of staple crops, and so unlike an ethanol economy it wouldn't compete with food production. The amount of methanol that can be produced from biomass through gasification is much greater than the ethanol that comes from biomass fermentation. Methanol production could be ancillary to other forms of power production and employed as a form of carbon capture. In this system, nuclear power could be used to split water and provide the hydrogen, while carbon dioxide could be captured from coal-burning power plants located on the same site. By burning coal, the carbon within it is used twice: the first time centrally when it's burned to generate heat (which is used to produce electricity at a power plant) and the second time locally where it's released into the atmosphere as carbon dioxide. By then capturing and converting the carbon dioxide into methanol, we can reduce the carbon intensity of coal by a factor of four (two from the co-location of the power plants, and two from the thermal composition of methanol). The same scheme can also be used for fracked natural gas, where, again, it would reduce the carbon intensity by a factor of four. Using fossil fuels more efficiently in this way would only be a stop-gap. In time we would be able to replace them with biomass, which would provide a carbon-neutral source of carbon dioxide, while the hydrogen would come from water split either by direct sunshine or electrolysis driven by solar PV.

This sounds almost too good to be true, but the first steps to achieving this vision are already taking place. Since 2006, China has rapidly established a domestic industry of coal-based methanol and dimethyl ether that is producing fuel at a price highly competitive with petroleum-based fuels. The methanol-based fuels offer many advantages over their petroleum counterparts, including a high octane rating and cleaner-burning properties. It's by no means perfect: a coal-based methanol economy would almost certainly exacerbate China's endemic water shortages, increase net carbon dioxide emissions, and add volatility to global coal markets; but all these problems can be addressed. As such, its expansion provides a tangible example of what could happen if George Olah's proposals become widely adopted elsewhere.

Perhaps the biggest problem faced by methanol and algal bio-fuel is that of traction. The issue of liquid fuels and petroleum replacements is typically subjected to hype and hysteria – it's difficult to separate fact from fiction. In this area, plausible science needs to be covalently bonded to sound economics, and reporters could do worse than brush up their understanding of the laws of thermodynamics. In October 2012, Air Fuel Synthesis (AFS), a small British company based in Stockton-on-Tees, announced that it had produced 'petrol from air'. As is fairly typical for this kind of announcement, blanket media coverage ensued. The *Independent* newspaper carried a report that epitomises the tone. It talked of a 'revolutionary technology that promises to solve the energy crisis as well as helping to curb global warming by removing carbon dioxide from the atmosphere', supported with this quote from Tim Fox, head of energy and the environment at the Institution of Mechanical Engineers in London:

> It sounds too good to be true, but it is true. They are doing it and I've been up there myself and seen it. The innovation is that they have made it happen as a process. It's a small pilot plant capturing air and extracting carbon dioxide from it based on well known principles. It uses well-known and well-established components but what is exciting is that

they have put the whole thing together and shown that it can work.

The company's chief executive, Peter Harrison, was happy to explain what all the fuss was about

> We've taken carbon dioxide from air and hydrogen from water and turned these elements into petrol. There's nobody else doing it in this country or indeed overseas as far as we know. It looks and smells like petrol but it's a much cleaner and clearer product than petrol derived from fossil oil. We don't have any of the additives and nasty bits found in conventional petrol, and yet our fuel can be used in existing engines. It means that people could go on to a garage forecourt and put our product into their car without having to install batteries or adapt the vehicle for fuel cells or having hydrogen tanks fitted. It means that the existing infrastructure for transport can be used.

Tim Fox is right: it does sound too good to be true. So is it true?

The chemical process employed by AFS removes carbon dioxide from the air by blowing it through a mist of sodium hydroxide, which converts it into sodium carbonate. The carbon dioxide is then removed from the sodium carbonate using electrolysis. It's then converted into carbon monoxide using a process called the reverse water-gas shift reaction, and finally combined with hydrogen gas, which has been split from water also using electrolysis, and converted into petroleum using the Fischer–Tropsch process. The science is entirely plausible. Each of these steps is a well-understood chemical reaction, and together they are very similar to a process for turning water into hydrocarbon fuels that was patented by the US government in 2008.

The problems are not with the science but the economics. Consider the three laws of thermodynamics, helpfully reworded here by Allen Ginsberg:

1. You can't win.
2. You can't break even.
3. You can't get out of the game.

In the AFS technique and similar processes, high-grade energy – electricity – is simply being converted into a source of low-grade heat energy, with significant losses in energy taking place every step of the way. In the AFS process, 6.7J of electric energy is used up in storing 1.0J of thermal energy in the petrol. In motor vehicles, the conversion of petroleum's heat energy into kinetic mechanical energy is in the region of 15 per cent; in an electric car, it's about 80 per cent. Consequently, the synthetic petroleum-powered vehicle will require about 35 times more energy for any given trip than an electric car and, one can expect, be about 35 times more expensive to run. Moreover, to synthesise enough petroleum to meet the current global demand of 89 million barrels per day would require around half of all the energy – in any form – being used globally. In comparison, petroleum represents just 6.5 per cent of global energy consumption. When all things are considered, AFS's synthetic gasoline looks like a non-starter.[1]

There's enough to be positive about in the prospects for algal biofuel, and the possibilities for methanol in particular, without looking for new ways to pull liquid fuel out of thin air. The bottom line with all the carbon dioxide-based technologies is that we need a cheap and readily available source of energy to turn carbon dioxide into a liquid fuel. Carbon dioxide makes up only a tiny proportion of the air – little more than one third of a per cent – and it needs to react with hydrogen before it can be turned into a fuel. Hydrogen might be plentiful, but it doesn't exist independently and

---

[1] Many thanks to technology writer Brian Dodson for his erudite deconstruction of AFS's Synthetic Gasoline Process.

all the techniques we have for liberating it from its chemical bonds are expensive and energy-intensive. Plants get around this problem by using sunlight to split water into free hydrogen and excreting oxygen via their stomata. The free hydrogen they create is stored by combining it with carbon dioxide to make carbohydrate, as we've seen. The methanol economy works on exactly the same principle – it uses carbon dioxide to liquefy the energy content of hydrogen, and it's why we need solar fuels and synthetic leaves. Even methanol from carbon dioxide, or any other hydrocarbon for that matter, makes sense only if we have vast amounts of cheap energy (potentially from solar, nuclear or excess renewable energy that would otherwise be wasted) to generate hydrogen.

Methanol should be thought of as a storage medium to conserve energy for future use; just like a battery in fact, but with all the added convenience of a liquid fuel. There are also many examples of direct methanol fuel cells, where methanol is oxidised within an electrochemical cell to produce electricity rather than heat. Making methanol in this way would allow us to store solar power for later and convenient conversion into electrical energy.

☀

There are many cases where liquid storage is not necessary – domestic heating, for example – and here there are some exciting technologies emerging. Cryogenic storage works on a similar principle to hydro-electricity. At night, when electricity is cheaper, it can be used to cool air from the atmosphere to the point where it liquefies (–195°C), taking up just a thousandth of the volume. During periods of peak demand, the liquid air is pumped at high pressure into a heat exchanger, which acts like a boiler but with air rather than water. Ambient temperature or hot water is used to turn the air back into a gas, and the massive increase in volume that ensues is used to drive a turbine.

In January 2012, the UK's first cryogenic power storage plant opened at Slough. This pilot scheme was operated by Highview Power Storage, which believes it has developed a relatively cheap way of storing power from intermittent sources such as wind turbines. If the pilot is successful, we may be much closer to being able to match the supply of renewable-powered electricity to demand, with the cost of storage predicted to be less than $1,000 per kW once the technology matures. This represents a cost less than one quarter of sodium-sulphur batteries and around a third that of pumping water uphill into reservoirs to deliver hydro-power. The technology is potentially ideal for places remote from the mountains and lakes required for hydro-electric schemes, like London and the densely populated south-east of England.

Highview has employed existing technology to build a liquefaction plant and power generator, which evaporates liquid nitrogen using waste heat from an adjacent biomass power plant. The process can also use ambient heat. Cold energy is then captured using a specially designed cold buffer employing a variant of the technology used in the steel and chemical industries, where cold is often stored in beds of sand or gravel. This cold energy can then be used to re-liquefy the nitrogen when it's next needed to store more energy, doubling the efficiency of electricity generation. The system can be applied to harness waste heat – and specifically low-grade waste heat such as that produced by the IT equipment in data centres – and returns about 50 per cent of the energy put in. This could rise to 70 per cent if waste heat from another source, such as a power station, is used instead.

Cryogenic storage is similar to the efficiency of much less energy-dense compressed-air storage plants, and compares to 70–85 per cent efficiency for batteries and 65–75 per cent for pumped hydro. The difference is the opportunity for scale. There's no reason why the system could not be scaled up to store and release hundreds of megawatts as a way of saving surplus electricity produced off-peak. And again, there are obvious benefits for using cold to store excess energy from wind and wave plants.

There's plenty of research taking place into energy storage in other areas. For example, flywheels are being used to store energy generated by wind turbines during off-peak periods or during high wind speeds. (In 2010, Beacon Power began testing a flywheel energy-storage system in Tehachapi, California.) Meanwhile, capacitors, devices similar to batteries that discharge their energy very quickly, are being used to supply energy to robots. The impact of these and other niche technologies is unlikely to be significant. However, the ability to store energy in a sustainable medium, using any of the technologies we have discussed, will have a significant effect on the way that energy is distributed and monetised. To some extent, we can expect that end-users will also become energy-producers. One example would be a building with its own turbine and PV system on the roof, with a battery set-up in the cellar. The University of Sheffield is considering the installation of such a combined heat and power system to ensure business continuity (and to keep Project Sunshine's research going if the national grid fails).

This kind of storage could be very local indeed. Today, most countries generate electricity in large centralised facilities, and while these plants have excellent economies of scale, the long distances over which they need to transmit electricity means that a lot of the energy is lost in transit. Local, small-scale generation allows collection of energy from many disparate sources, offering greater efficiency because the energy is produced near to the point of use, and potentially greater security as well, by reducing the number of power lines that need to be constructed and maintained. Historically, these benefits needed dedicated operating systems together with expensive and complex plant to minimise pollution, but embedded systems have been developed that can provide them with automated operation using renewable power sources. These small-scale systems can generate anything from 3 kW and 10,000 kW. The most popular systems use PV panels and generate electricity at a price comparable to coal-powered plants. There are some technical issues surrounding power quality, but in principle the energy produced can be integrated into a central grid to provide greater reliability.

A microgrid is a localised collection of electricity generation and storage, operating independently, that is also connected to a centralised grid. From the operator's perspective, a microgrid can be treated as a single entity. The ability to isolate individual microgrids allows a high level of control and provides a reliable supply of electricity from a disparate number of sources. There is an increasing frequency of blackouts and brownouts in electrical grid systems. Some are caused by operator error, others by equipment failure, yet more by extreme weather events that either create unattainable peaks in demand or bring down facilities; and even electrical storms on the sun can affect grid performance. Microgrids were proposed as a solution to the problem that caused a major blackout across many regions of India in July 2012.

It's clear that population growth is going to be the primary driver of demand for energy. There is a plausible case to be made for delivering a large proportion of that energy using carbon-free technology, and given time it's certainly true that all our energy needs could be met exclusively by renewable sources, once we've overcome the problem of intermittency. While this should be of some comfort (not least because once the oil runs out – whenever that might be – we won't have an alternative), the timescales we are required to operate within, and the scope of the research and engineering involved are truly daunting. Yet, while the challenge is considerable, we have no option but to rise to meet it.

Considering only the technical aspects of energy supply, it's entirely reasonable to conclude that we are well equipped to solve the energy crisis. A plausible scenario for a sustainable energy future is one where there's a global agreement that we ramp up investment in solar photosynthesis, increase nuclear capacity, capitalise as much as we can on wind, wave and solar PV, while at the same time reducing fossil fuel consumption, using carbon capture and replacing petroleum with methanol and algal biofuel from environmentally friendly sources. This is easy to imagine, but just because it could happen, that doesn't mean that it will. Available technology is not the only driver in the energy markets: vested interest, demand, resources and to a lesser extent the environment are all key considerations.

An alternative, equally plausible scenario, identified by the energy companies, is one where the concerted effort and planning required to achieve sustainability fails to take place. If demand-side policy is not pro-actively pursued until the supply limitations start to hurt, there will be a scramble for remaining resources driven by national self-interest. Environmental policy would not be addressed until the effects of climate change became too serious to ignore (by which time it might be too late to do anything about them). Instead of international cooperation, there would be a network of bilateral agreements, as governments ignore the consequences of failing to curb the demand for energy in order to maintain popularity. This pragmatism would lead to an environmentally disastrous plunge into coal, and a piecemeal approach to biofuel, renewables and solar that would make the achievement of a sustainable global energy economy extremely difficult, if not impossible, to achieve. One of the first casualties of the economic crash in 2008 was the raft of energy policies designed to combat global warming by reducing carbon emissions and funding renewable energy research. Many politicians claimed at the time – and still do – that the road to economic recovery required the abandonment of sustainable energy in favour of pragmatic activities, undoubtedly cheaper in the short term, such as fracking for natural gas.

There is no perfect solution to the problems we face: all of them have drawbacks. Despite the pressures on our energy system, its complexity and scale mean that there's a great deal of inertia within it, which is exacerbated by even mild resistance to change. Policies made today will not pay dividends for 30 years, and we need to accept that. Our politicians and business leaders need to be brave. And so do our scientists.

As Jeremy Grantham, founder of the Grantham Institute for Climate Change at Imperial College London, says:

> It is crucial that scientists take more career risks and sound a more realistic, more desperate, note on the global-warming problem. Younger scientists are obsessed by thoughts of tenure, so it is probably up to older, senior and retired scientists to do the heavy lifting. Be arrested if necessary. This is not only the crisis of your lives – it is also the crisis of our species' existence. I implore you to be brave.

There are some choices to be made, but they need to be made quickly. We could bring all the solutions online much, much earlier with more funding. Solar power really is the future, but funding into the field is pitifully small. The US government spends $28 billion on health, but only $28 million on basic solar research; more is spent at the nation's gas stations in a single hour. If we decide that we're serious about addressing our energy problem there's plenty of evidence to make us optimistic about succeeding. In terms of figuring out a way of mass-producing a liquid fuel from solar energy – methanol – that can be pumped, piped and poured, we aren't taking shots in the dark. This isn't like fusing atomic nuclei: solar power has been with us for decades, we know how to capture and store sunlight, while plants provide us with a highly visible proof of concept for photosynthesis. Solar can provide the planet with all the energy it needs: no other energy source has that potential. We all need to get behind it.

14 Feast or Famine?

> 'Farming looks mighty easy when your plough is a
> pencil and you're a thousand miles from the corn field.'
> *Dwight D. Eisenhower, US President, 1953–61*

> 'Never eat more than you can lift.'
> *Miss Piggy, Muppet*

Food is now so pervasive that we tend to take it entirely for
granted. An omnipresent fixture in our lives, it accounts for
almost 30 per cent of all TV advertising. A medium-sized
supermarket stocks approximately 45,000 products, allowing us to cram our larders and fridges with exotic produce
from all over the world. And eating out is the most popular
recreation activity in the UK and the US, ahead of going
to the movies or watching and playing sport – it even beats
shopping into second place. We can eat whatever we want,
which is what a lot of us do. A third of children in the UK
and USA are overweight or clinically obese. In the UK,
44 per cent of adult males and 33 per cent of women were
classified as overweight in 2009.

With obesity the overriding public health issue in the
developed world, it's easy to forget that food hasn't always
been so readily and cheaply available. Little more than 30 years
ago the typical UK diet was very different to that of today,
but previously it had remained relatively unchanged for decades. Before 1980 much less meat was consumed; offal was a
central constituent of many home-cooked meals and a staple
of the school canteen. We ate less rice, pasta and processed
food (which was expensive) and, proportionally, more potatoes and green vegetables. In 1980 only 6 per cent of men
and 8 per cent of women were obese. We ate much the same
food in much the same way as our parents and grandparents.
Three generations, all old enough to remember a time when
food was proportionately much more expensive (accounting

for a third of household income during the 1950s, compared with just 17 per cent today) but young enough to have benefited from the Haber–Bosch process and the ensuing green revolution, which gave us affordably full stomachs.

But this bounty has not been evenly shared. Today at least 1 billion people worldwide are chronically undernourished; 180 million children are severely underweight; 400 million women are suffering from anaemia. This is due entirely to a shortage of capital at a local level, not to a worldwide shortage of food. In Africa, food intake per capita is 20 per cent less than it was in 1960. Ironically, this fact would be a good news story if it was about western Europe or the USA, but in Africa it simply means endemic starvation and widespread illness. Here in the West, the years of seemingly endless harvest have made us complacent. This is dangerous, because for six out of the last eleven years we have not been producing as much as we have consumed. It may not seem like it, but we are five years into a global food crisis that is unlikely to disappear any time soon, threatening poor countries with increased malnutrition, starvation and even total economic collapse. In the West we notice that the price of food is increasing, and this trend is forecast to continue, with many economists projecting a two-fold rise by 2030. This will put the cost of the weekly shop at levels not witnessed since the early 1950s.

There were many reports in the autumn of 2012 that the year's extreme weather had impacted on the harvest in many of the major food-producing states such as the Ukraine and the USA. It was, after all, a record year in the US for droughts and heatwaves. In these countries and elsewhere, falling harvests were blamed for the erosion of grain reserves, which globally stood at their lowest levels since 1974, at an average of 74 days of consumption, compared to 107 days a decade earlier. In October 2012, the price of wheat and maize stood close to the peak price that had led to riots in 28 nations during 2008. At the same time Lester Brown, the president of the Earth Policy Institute in Washington, warned that climate change meant the weather is now so unreliable, and the demand for food growing so quickly, that unless urgent action is taken the collapse of the global food supply system is inevitable, leaving hundreds of millions more people without enough to eat.

In nature, there are many examples that show what happens when a glut of food comes to an end. Mautam is an extreme ecological phenomenon that occurs in Mizoram and Manipur, two states in north-eastern India. Approximately a third of this area is bamboo forest and once every 48 years a particular species, *Melocanna baccifera*, flowers at exactly the same time across the whole area. After flowering, the plant dies and regenerates from its seeds, which are superabundant at this time. The chief beneficiaries of this rare harvest are the local black rat population, whose numbers soar as a consequence. Once the rats have exhausted this initial food supply, they leave the forests in their millions and turn their attention to cultivated crops and food stores. These so-called 'rat floods' cause a year-long famine. In 1862 and 1911 records taken from the British Raj show that severe famine was experienced after the bamboo flowering. People talked of rats 'suddenly and magically' appearing in plagues, eating everything they could. During the ensuing famine, the rat population collapses.

Just like the rat floods of mautam, we can live well through the good times, but they can't go on for ever. There is a general assumption that to provide sufficient calories for 9 billion people we need to increase food production by between 60 per cent and 100 per cent. We also need to produce much more meat to satisfy the developing world's burgeoning middle classes. The scale of this challenge is dangerously underestimated not only by the public and the media, but also by governments and financial markets: the very people we look to for appropriate action if we are to have any chance of meeting this target. The energy situation we have outlined in the previous chapters is no more pressing in the short term than the challenge of achieving food security. In fact, oil is so entwined with food production, from fertiliser through to logistics, that it's impossible to separate the two. If the price of oil goes up, so does the price of food; and a shortage of oil means less food can be produced.

The impact of these problems means that in rich countries, the price of food will continue to increase and national budgets will be mercilessly squeezed, hitting profit margins and stifling economic growth. In poorer countries hundreds of millions of people will starve to death.

Given the scale of the crises, it would be remiss to suggest that there's a quick-fix solution. As the UK government's chief science advisor, Sir John Beddington, says: 'There will be no silver bullet.' And yet there remain serious grounds for optimism. There are technologies available that can alleviate many of the worst dangers that lie in the years ahead. Surely it would be foolish not to make the maximum use of them?

※

Artificial selection within plant species is as old as agriculture itself. There are even references to the importance of selective plant-breeding in ancient texts by Virgil and Aristotle. The earliest farmers selected species on the basis not only of how well they grew, but also on their hardiness and resistance to pests and disease. The choices they made formed the basis of our modern staples – wheat, barley, lentils, rice, maize and others – which are descended from those original strains.

Classical plant-breeding involves deliberately cross-breeding closely related varieties of one species to produce a new strain with the beneficial characteristics of both parents. This new strain is a hybrid of both, and so this process is known as 'hybridisation'. Prior to the rediscovery of the work of Gregor Mendel in the early 1900s, hybridisation was an exercise in trial and error. With no knowledge of genetics, farmers were unable to understand why some characteristics would disappear for one or more generations only to reappear, apparently at random, some point further down the line.

The mechanism of recessive genes and the many other findings of Mendel gave plant-breeders an insight into the nature of trait inheritance, which they could apply through deliberate pollination to produce predictable characteristics in hybrids. These techniques led to a significant increase in crop yields, particularly in the United States, which in turn led to further methods of developing better crops during the period up to the Second World War; a foreshadowing of the massive yields that would later be enjoyed by cereal-breeders throughout the green revolution. Today, yield increases are marginal, if not static. Many breeders believe that this plateau in production is a function of the degradation of soils in agro-ecosystems. Whatever the reason, it's clear that another step-change in cereal yields is required.

The technology with the greatest potential to help us achieve this is the increased use of genetically modified organisms (GMs). Take the issue that currently just under a third of all food is lost prior to harvest, usually eaten by pests that human beings have failed to control, or because of water shortages or the increasing salination of fresh water supplies. These are exactly the kind of problems that biological engineering is best placed to address. For example, the pests. Each year in the UK alone, aphids cause around £100 million of damage to agriculture. Scientists working at Rothamsted research station have engineered a strain of wheat that discharges a chemical called E-beta-farnesene, the same as that produced by aphids when they are threatened. It also attracts aphid predators such as wasps and ladybirds, which are harmless to the wheat. The result is that the aphids either fly away or are eaten before they can do any damage.

In the past, field trials of GM crops have been regularly threatened by anti-GM campaigners. But with around 1 billion people already suffering from serious food shortages and starvation, it's inconceivable that we will not face a much bigger problem over the coming years unless we take action now and explore the range of possible solutions.

Despite all the benefits, you'll rarely find GMs enjoying

positive coverage in the media. They are typically presented as an unnecessary option. Many critics of biotechnology and technologically advanced agriculture support organic farming instead. It's often claimed that organic produce is healthier, more nutritious, tastier and better for the environment. The Rodale Institute in Pennsylvania has been running an organic farming longitudinal experiment for more than 50 years. It's the longest-running side-by-side US study comparing conventional chemical agriculture with organic methods, and indeed it has found that organic yields can be made to match conventional and out-perform them in years of drought and environmental distress. Rodale tells a good story, but unfortunately what they don't say is how much human effort is involved in making this model work. Their farm is staffed by many more workers per hectare than typical farms and these workers are not typical farm labourers either: they are very well educated and also committed to a cause. In fact, despite half a century of comparative research, scientists have been unable to find any evidence that organic foods are healthier than non-organic, and blind tastings invariably find that people can't tell the difference.

Even the notion that organic farming is more environmentally sound than conventional agriculture is questionable. Organic farmers do not use synthetic pesticides, it's true, but they still make use of chemicals that can be ecologically damaging. Natural manure contains high levels of nitrates that pollute the air and the water. Perhaps most tellingly of all, the majority of studies have found that, on a large scale, and with comparable manpower, organic yields are at best only around 80 per cent of those achieved by non-organic practices. One of the biggest issues is the use of so-called 'organic pesticides'. In the past, the only way of controlling fungal pathogens in organic farming systems was to spray them with copper sulphate. Its use is especially prevalent in organic potato farming, but the environmental contamination associated with using metals in this fashion is enormous. In humans, copper sulphate is an irritant, moderately toxic if ingested, but copper itself is poisonous to plants in high doses and extremely harmful to fish.

Although it's categorically not the answer, it would be wrong to dismiss organic farming out of hand – there's a lot that conventional agriculture can learn from it. Indeed, the question of whether organic farming can feed the world is somewhat fatuous: we are where we are. The Haber–Bosch process to make artificial fertiliser has led directly to the Earth's population quadrupling in less than a century. Today, 5 billion people depend existentially on food grown in soil that has been artificially fertilised. Even if it were possible for the whole world to go organic, which would require the appropriation of more land and more water, we could feed only around 2 billion people because the yields would be much lower. However, organic demands a much closer relationship with the soil and a better understanding of the microbes and fungi within it. By naturally enhancing soil fertility and increasing the biodiversity found in plots, organic farming is less dependent on external inputs. So although crop yields are 20 per cent lower, these have been achieved using 34 per cent less energy and fertiliser and 97 per cent less pesticide.

The issue of organic versus conventional highlights the biggest challenge for developing a global strategy for agriculture: the present culture of confrontation not cooperation. The Hudson Institute is a conservative not-for-profit think-tank based in the US. The Institute is supported by big businesses including several of the oil companies, so perhaps unsurprisingly it's highly critical of environmentalism and organic agricultural methods. Its claims, if not always at odds with the scientific community, can be misconstrued, such as the statement by Dennis T. Avery, Hudson's head of global food issues, made in January 2010, that 'Higher CO_2 levels should mean higher crop and livestock yields! Tally ho!' While it is indeed true that higher levels of CO_2 should stimulate photosynthesis in certain plants (but not low-latitude crops such as maize, sorghum, sugar cane and millet, plus many pasture and forage grasses), any benefit would be wiped out by the increase in temperatures. Mid-latitude yields would also be reduced by as much as 30 per cent due to summer dryness.

※

Climate change is one of the most used and abused issues when it comes to agriculture. Despite the more obvious targets of long-haul flights to exotic holidays and 4×4s, the meat in our diet is one of the primary sources of greenhouse gas (GHG). Sheep and cows are ruminants and their digestive system produces a lot of methane – the source of that familiar 'farmyard' smell. Methane is a GHG with around 20 times more global-warming power than carbon dioxide. Pigs produce less methane than cows, but plenty of manure, while chickens eat and waste very little in comparison. The efficiency with which each of these animals converts vegetable matter into meat protein is also vastly different. Cows and sheep need 8 kilos of grain to produce 1 kilo of meat, pigs require about half that amount, but chickens need only 1.6 kg. A further consideration is the amount of land you need: the more required to feed each animal, the less you have left for anything else. When you take all this into account, you can't help but end up with some frankly unpalatable conclusions. Indeed, in 2008, the UN's food and agriculture organisation weighed up the evidence and determined that livestock were warming the planet more than transport, which in turn prompted the Vegetarian Society to launch a campaign from the moral high ground declaring that our 'silent but deadly' carnivorous diet was responsible for destroying the climate.

As we have witnessed with the behaviour of the oil companies, big business tends not to be particularly receptive to environmental challenges, especially those that threaten to erode their market share, and businesses don't come much bigger than the food corporations. On one level, then, their response can be considered admirable. Fearing that low-carbon tyranny might drive them out of business, they chose not to attack the science but rather to investigate low-GHG livestock solutions. Unfortunately for the animals, many of these were to be found indoors. Housed animals are easier to manage, require less space, and don't waste energy running around enjoying themselves. The manure they produce is harvested and burned as carbon-neutral fuel, thereby avoiding most of the harmful impacts on the environment. Plans are for sealed barns to have exhaust vents in which GHG can be captured before it enters the atmosphere.

From a purely environmental perspective the numbers stack up. Using these criteria, indoor-reared poultry is the greenest meat you can eat. Organic free-range poultry has about 45 per cent more global-warming potential. But before the vegetarians get any smugger, there's also the issue of dairy, because most of those who avoid meat are not vegan and source much of their protein from foods made by animals pumping out GHG. So if you are avoiding meat for climatic reasons, you should probably think about steering clear of dairy as well. There are options for getting GHG-free dairy, but they too involve intensive indoor husbandry. If you're cold enough to ignore the fact that we're dealing with advanced life-forms, i.e. sentient, living things, then we suppose this is definitely an option.

To achieve the economies of scale we require to feed 9 billion or more people, we'll have to make some pretty unpleasant decisions. The bottom line is that to meet this demand we simply cannot get away from industrial agriculture, whether we want to or not. An exclusively organic, free-range, GM-free food production is a recipe that might feed 900 million people, not 9 billion.

The methods of industrial farming are techno-scientific, political and economic, including mechanical innovation, genetic technology and improvements in the processes of farming. These techniques are applied to maximise yields, they are practised throughout the developed world, and they are responsible for producing almost all of the meat, dairy, eggs, fruit and vegetables we consume. Global agricultural production doubled four times between 1820 and 1975, supporting 1 billion people at the beginning of this period and 7 billion people today. However, during the same period the proportion of the population involved in agriculture has declined massively. As recently as the 1930s a quarter of the American population worked in farming, but today the sector employs just 1.5 per cent. Industrialisation has, however, led to a concentration of farm ownership. Just four companies slaughter 81 per cent of cows, 73 per cent of sheep and over half of all chickens and pigs in the USA. This vertical integration has provided us with enough food to eat, but notwithstanding the significant moral issue of animal welfare, there is some scepticism about whether it can continue to do so.

Currently worldwide there are around 1 billion pigs and sheep, 1 billion cows and more than 50 billion chickens. The biomass of livestock already exceeds that of mammals and birds living in the wild, and by 2050, if we don't alter our diets and we continue down our current path, we will require twice as many cows, sheep, chickens and pigs to satisfy both the increased population and the greater demand from the burgeoning middle classes in the BRICs.

The first question is what are all these animals going to eat? Returning to the reports about the harvest of 2012, it was not, as was widely reported, one of the worst global harvests in years. Despite the extreme weather, it was one of the best: just 2.6 per cent down on 2011, the highest on record. The problem is that the diversion of so much grain, to feed animals and for conversion into biofuel, together with the rising population, means that we need a record harvest, year in year out, every year, come rain, wind or snow.

The harvest of 2012 was the third-biggest on record, yet it was still not enough to avoid a food deficit, and the market consumed 28 million tonnes more grain than the farmers produced. We must do more than hope for a bumper harvest in 2013 and every year afterwards. Improvements in grain productivity have slowed year on year since 1970: from 3.5 per cent per decade to just 1.5 per cent. It's reasonable to assume that we have hit a glass ceiling as far as existing technology is concerned. Our current grain species are at or nearing their biological limit (interestingly, over the same period, funding for agricultural research has steadily fallen, as a per cent of GDP, across the world). Some people are suggesting rather radical solutions to this problem. The carbon footprint of those most beloved of family pets – cats and dogs – is huge. A single cat has the same environmental impact as a Volkswagen Golf driven 6,000 miles each year; a medium-sized dog is equivalent to a 4×4. Would we be prepared to do away with our furry friends to ensure that we can enjoy meat with our two veg? Some argue that whatever we think, we will be forced to go even further: all become strict vegans and make our last meat meal our pets!

Thankfully there are some alternatives to this rather extreme strategy, but before we can say definitively what we are going to be able to eat, we need to consider how food is not only produced but distributed. Food and energy are inextricably linked. 'Food miles' as a concept originated in the 1990s, referring to the distance that food is transported between the point of production and consumption. The concept is not perfect – food miles aren't always correlated with the actual environmental impact of food production – but they serve our purposes for generally highlighting the environmental impact of food distribution on global warming. The average number of food miles for almost all foodstuffs has increased significantly over the past 100 years. This is due to several factors including: the globalisation of trade; the consolidation of food production into fewer, larger regions of the planet; and the increase in the consumption of processed food, which necessarily makes a number of detours on its journey from farm to plate. 83 per cent of GHG emissions associated with food occur during these production phases. Even if we do produce enough food globally to feed everyone satisfactorily, the continual rise in the cost of production due to the increasing price of oil will mean that many people simply will not be able to afford to eat. If we don't, and food pressures do recur, due to a failure to increase output or the impact of climate change on the harvest, and these are reinforced by an increase in the price of oil, then the risk of civil unrest and even social collapse will increase to a point where they will probably become the major source of international confrontations.

To avoid either of these scenarios our entire agricultural system needs an overhaul. A holistic 'systems approach' is required: we need to view the whole world as a single collective farm and develop a global agricultural policy. All agricultural systems comprise four sub-systems: agriculture and land strategy; production and harvesting; processing and distribution; and purchasing and consumption. We have to maximise efficiencies across all four if we are to avoid the unpalatable decision of either letting half the world starve or looking forward to a diet of beans and pulses (after a farewell-to-meat beano, featuring Fido as the main course).

There is much to be done, but there is also much that we can do.

☀

Part of the problem with food supply is that it's largely governed by a small number of mega-corporations and mega-corporations tend to be driven exclusively by market forces. They are usually robust in their resistance to regulation in principle and practice. However, food is simply another form of energy, and developing a global strategy should be the collective responsibility of sovereign governments. This should not be misconstrued as a call for nationalisation, but rather for true globalisation and the development of an international strategy that will work for everyone – regardless of where they happen to have been born. There is a major role for the mega-corporations to play: we are not against the profit motive as such, rather the notion that it should be the only motivation. Food companies should be operating within a framework working towards an achievable goal, rather than operating outside it as they are at the moment.

Nor should we forget the important yet usually over-looked role that farmers play. In the UK, migration from the countryside to the cities means that domestic food production involves just 2 per cent of the population working on the farm. This 2 per cent, who essentially provide the food for the remaining 98 per cent, are among the most poorly-paid workers in society. Mean earnings in the sector are just £15,000 a year, less than two thirds the national average, hours are long and working conditions less than salubrious.

Getting the supply-side economics right is of paramount importance. Taking a global view would allow key decisions to be taken about land use (agronomics) at a strategic level and the choice of crop species and genotypes could be influenced by offering economic incentives and subsidies. This would ensure the kind of harvest the market needs, rather than the one that the market simply wants.

There is a great disparity between the yield from farms in the developed world that have benefited from industrialisation and those in the developing world that haven't. Introducing the technology and practices that are regarded as standard in Europe and the US to farms across Africa and large parts of Asia will make a big difference to productivity. There are high-tech farming projects already under way in desert regions of Australia, the Middle East and the Canary Islands, which show what can be achieved. In these areas, the only abundant natural resource is sunshine. Sundrop Farms, for example, has operations in South Australia and Qatar in which it's growing high-quality pesticide-free vegetables all year round in commercial quantities, using solar power to desalinate seawater for irrigation and to regulate the temperature within its greenhouses. The endeavour uses no fresh water at all and hardly any fossil fuels. The success of the eighteen-month project has been so great that the company has already secured £8 million in funding for an 8-hectare greenhouse that is 40 times larger than the pilot scheme. On completion, the facility will produce 4 million tonnes of tomatoes and peppers each year. Given that traditional agriculture accounts for 60–80 per cent of all fresh water supplies and 40 per cent of all land, producing such bounty using barren desert and no water at all makes this commercially viable project arguably one of the most exciting currently happening anywhere in the world.

The farms use a system of motorised parabolic mirrors, which track the sun during the hours of insolation and store the heat in a pipe containing a sealed supply of oil. Seawater is pumped from the ground and superheated by the pipes to a temperature of 160°C. The steam is then used to drive turbines and produce electricity. Some of the hot water provides heat in the greenhouse during the cold desert nights, while the rest is converted into 10,000 litres of fresh water each day at a desalination plant. The system uses the latest technology – the farmer can even moderate conditions in the greenhouse via an app on his iPhone – but it's commercially and environmentally scaleable: there is, after all, no shortage of sunshine or salt water.

Technology can also help to make traditional agriculture more sustainable and productive. Here we should begin by looking at the earth. The soil, or till, forms the heart of the world's critical zone: a thin veneer that extends from the bottom of the underwater aquifers and artesian wells to the top of the tree canopy. Soil is formed as rocks break up and are dissolved by micro-organisms to form particles that bind with decaying biomass to form larger particles. These aggregates provide a good balance of nutrients and minerals for plants to feed on, but they also facilitate drainage and are loose enough to allow oxygen to reach the roots. This natural capital that builds up is vital to life, but also extremely fragile. It can be flushed away by the rain or blown away by the wind. It's also degraded by pollutants or salts that build up after irrigation water has evaporated, and compacted by heavy machinery. Much of it is sealed up beneath our towns and cities. In Europe, the urban geographical footprint has increased by almost 80 per cent since the 1950s, and the coming decades will see similar levels of depletion elsewhere as the process of urbanisation continues to be driven by population growth. The soil is also sensitive to climate change. In warmer conditions, the microbes within it can degrade organic matter much faster, which releases carbon dioxide and other GHGs into the atmosphere and depletes the desired aggregates. A recent study found that soil in England and Wales is losing 0.6 per cent of its carbon content each year.

The best way to look after the soil is to allow the soil to look after itself, and that means adopting some of the soil-management strategies practised by the organic movement. Current farming is essentially an alfresco version of hydroponics: cultivation in a sterile environment. The natural microbes are removed from the till and nutrients have to be artificially introduced. Fertiliser, pesticide and herbicide are expensive, while their production and distribution is dependent on oil. Sterilising the soil in this way does allow a great deal of control, particularly when it comes to managing pests, but it's not the most sensible way to achieve it. The problem is that not all the microbes are harmful to the plant – in fact many of them are highly beneficial and we are only just beginning to learn how these relationships caused the world to go green in the first place.

A 2012 study by the University of Sheffield has shown how fungi dwelling in the soil affected the evolution of plants. Scientists recreated environmental conditions from more than 400 million years ago in growth chambers and nurtured prehistoric species within them. This ground-breaking work is providing a fundamental understanding of how plants managed to colonise the land before they had evolved roots: they gained growth-promoting soil phosphorus from the fungi in exchange for sugars fixed by the plant through photosynthesis. The co-evolution of plants and arbuscular mycorrhizal (AM) fungi is one of the most ancient symbiotic relationships. It played a crucial role in the evolution of all the Earth's ecosystems, yet in the last century it has been all but lost to agriculture across many regions of the world.

AM fungi also perform an important role as ecosystem engineers. Symbiotic interactions between AM fungi and other microbes can lead to shifts in the structure and function of host plant communities. They can be used as a tool to suppress weeds and benefit the crop species. Viewed in this way, they represent a potential alternative to costly and environmentally damaging herbicides. Over several generations of rearing in this sterile environment, plants become entirely dependent on the artificial fertiliser and herbicide, effectively shunning the symbiotic help that AM fungi offer them, so getting the right combination of herbicides becomes critical if the crops are not to be overrun with weeds.

The act of ploughing fields is also an exercise in pest-control. However, it causes all sorts of problems for the AM fungi. If we are to breed plants that can take advantage of the AM fungi, then we need to drill rather than till. The plough may have served prehistoric farmers perfectly well, but there's no need for it today – much better to use a seed drill instead, which will enhance the soil quality. This kind of conservation tillage is practised on 45 million hectares worldwide, concentrated in North and South America. It's carried out primarily to protect the soil from compaction and erosion, but also to reduce the costs of production. Conservation tillage improves soil structure and stability, reducing the extremes of flooding and drought by increasing holding capacity and providing better drainage. There is also a reduced risk of run-off and pollution of surface water by pesticides, fertilisers and other sediment. There's more research to be done to see whether conservation tillage can offer the same kind of benefits elsewhere in the world. In Europe, for example, no detailed studies have been concluded so far, but we can be cautiously optimistic that conservation tillage might offer benefits across a range of climatic, cropping and soil combinations.

※

The debate about GM that dominates the mass media is

emblematic of the complacency towards the food supply that has become ingrained across the developed world. The fact is that GM is absolutely necessary to produce the yields that we require today – not something that we can muse over as an ethical dilemma for a future generation. Without GM people will starve in their tens, if not hundreds, of millions. The moral question is not whether we can afford GM, but whether we can afford to waste time having a debate about it. We should of course recognise that there's an issue with GM being the preserve of a few global mega-corporations, and that's something we need to address, but it would be folly to throw the baby out with the bath water.

The potential upside for GM is significant. Bio-engineering is well placed to address many of the problems we have outlined: salination; climate change; drought; flooding. It can lead to a significant reduction in the use of artificial pesticides and herbicides (by developing plants that can produce their own) and it can improve both crop yields and the fertility of the soil. GM can be applied effectively to all the major staples: maize/corn; rice; wheat; potato; cassava; soya bean; sweet potato; sorghum; and yam. Scientists are very optimistic that they will be able to engineer more efficient photo-synthesising 'C4' genes (from a family that includes corn, sugar cane and millet) into relatively inefficient but vital 'C3' plants such as rice and wheat within 20 to 30 years. Success in this one area would increase yields of C3 crops by up to 50 per cent. This would buy time (the very thing in shortest supply) and allow for a less painful transition to a stable and sustainable population.

This might sound suspiciously like science fiction, but work is already well under way. The C4 rice project, a 15–25-year programme aiming to modify how the plant photosynthesises to boost crop yields, has been ongoing since 2008. At the International Rice Research Institute (IRRI) in the Philippines, supported by the Bill and Melinda Gates Foundation, technical director Paul Quick and his team have been working on changing the photosynthesis process in rice from the C3 carbon-fixation mechanism, common to 98 per cent of all plants, to its much more efficient C4 counterpart. The plants that have evolved this process naturally – such as maize – are able to devote more of their energy to carbon-fixing, and thereby growth. Nourollah Ahmadi is a rice specialist based in France at the Centre for International Cooperation on Farming Research for Development: 'At present we are studying how far the cell structures and enzymes required to achieve C4-type photosynthesis are already present in rice and closely related plants. The aim is to activate the available but as yet inactive cell structures and enzymes and to introduce the ones that are lacking by genetic transformation, drawing on other plants.'

An increase in output by as much as 50 per cent would herald another green revolution, and certainly give us enough food to be capable of meeting projections for demand up to and beyond 2050. Rice is the staple foodstuff for more than 50 per cent of the world's population. In 2009, production stood at 678 million tonnes per year. For every additional 1 billion people, an extra 100 million tonnes of rice needs to be produced, but the annual growth in yields has slowed since the 1990s from 2 per cent to just 1 per cent per year as the species reaches its biological limit. In short, without GM there is no chance of us hitting these targets.

GM requires fewer inputs, but we are still going to be relying on Haber–Bosch for the foreseeable future to provide nitrogen. There also remains the question of where the phosphate is going to come from. There is uncertainty over the size of the world's phosphate reserves, compounded by the fact that supply is controlled by a small number of players. China, Morocco, the USA and Russia hold more than 70 per cent of the market, presenting the possibility for a global cartel to emerge. There is some evidence of strategic manoeuvring. In 2004, the USA and Morocco signed a free trade agreement covering the import and export of phosphate. Four years later, Morocco was exporting $65 million-worth of fertiliser to the USA. The deal was aimed at future-proofing the USA's food and fertiliser supply. Although it has one of the largest reserves of phosphate at the moment, production is expected to drop off in 25 years' time. Unlike the other finite resources we have discussed, there is no substitute for phosphates. Cutting usage will make a difference, but bigger gains will come from reprocessing and recycling it.

There is a lot of phosphate locked up in the soil, and one solution would be to find a key that allows plants to get hold of it. AM fungi could provide the means to achieve this, as they can access forms of phosphate that plants are unable to, or help them to exploit unlocked phosphates from a much greater volume of soil. The challenge is two-fold. Firstly, free phosphate in soils is locked up by geochemical processes. With metals, especially aluminium, this occurs first through a molecular bond called chelation and through subsequent binding with clay particles. Secondly, most phosphate in the soil is of a chemical form that is inaccessible to plants. A suite of adaptations allows plants to access some of these phosphates, including their own photochemistry, but micro-organisms, especially AM fungi and rhizobacteria, play a vital role in mineralising this phosphate and thereby liberating it for plant growth.

There is also a much richer source of phosphate: the manure generated by pork and dairy farming. Waste from livestock contains around five times more phosphate than human excrement and there is enormous potential for recovering this resource. There is a programme investigating how to resolve this, and if successful, the amount of phosphate fertiliser produced in the US alone would be enough to make it self-sufficient. This is one very good reason why we don't all need to turn vegan.

Sustainability has to be at the centre of all efforts to help provide the world with enough food. This will be achieved only by broadening the focus beyond individual plants and animals; by gaining an understanding of the complex relationships between species. Nurturing crops in sterile soils, relying on artificial interventions to meet the challenge of pests and plant nutrition, is a palliative rather than a sustainable solution. In the future our food supply must be able to withstand the challenges of pests and climate change. Here again, GM is helping us to develop crops that can themselves engineer beneficial symbiotic relationships that reduce their reliance on further artificial interventions with pesticide, herbicide and fertiliser.

This is another result of research being carried out in the UK at Rothamsted, in collaboration with the University of Sheffield. The team discovered that maize emits chemical signals that attract growth-promoting microbes to its roots, boosting resilience, growth and ultimately yield. Dr Andrew Neal, jointly leading the research, explains:

> We have known for a while that certain plants exude chemicals from their roots that attract other organisms to the area. In fact, the environment around a plant's roots teems with micro-organisms and populations of bacterial cells can be up to 100 times denser around roots than elsewhere.
>
> Simple compounds such as sugars and organic acids are attractive to these micro-organisms as they are a good source of energy; however other more complex chemicals were not known to serve as attractants because they were typically thought of as toxic.
>
> Now we have evidence that certain bacteria – we studied a common soil bacterium called *Pseudomonas putida* – use these chemical toxins to locate a plant's roots. The plant benefits from the presence of these bacteria because they make important nutrients like iron and phosphorous more available and help by competing against harmful bacteria around the root system.

The soil around the roots of every plant is awash with chemicals that have been released. This makes the soil rich in nutrients, but much more toxic for micro-organisms. The roots of young maize plants release large quantities of chemicals that play a role in helping the plant to defend itself through stem and leaves against pests above the ground. Work is now under way to obtain a molecular blueprint of the microbial communities that are shaped by these root chemicals, and to investigate what beneficial impacts these microbes have on plant growth, plant health and soil quality.

This kind of GM approach is simply a way of looking to the past for solutions to today's problems. The oldest root–fungus partnership has been found in the first trees from 350 million years ago. The relationship accelerates mineral alteration. Over time, the evolution of both the root-associating fungi and their tree partners accelerated the processes that drive soil development and release calcium from continental rocks into the oceans, locking up CO_2 in the mineral calcium carbonate. This formation of calcium carbonate contributes to the regulation of the Earth's carbon dioxide levels over timescales of millions of years.

✳

While there are clearly big improvements that can be made during production and harvest, arguably even these huge gains pale in comparison to the savings that can be made by improving the processing and distribution efficiency of our current output. Against the background of shortages, endemic hunger and escalating prices, it's a chilling fact that in the US and UK today, a third of all food produced is thrown away. Almost all of this waste is entirely avoidable, the result of befuddled logistics and consumer whimsy. Supermarkets know that shoppers like to see full shelves, so they ensure that they are full, whether you're shopping at 8.00am or 8.00pm. In practice, this is hugely wasteful. At the end of each day fresh bread, meat and dairy is turfed out by the skip-load. There is also the question of what food looks like. The beautiful presentation of fruit and vegetables, uniform in size, shape and colour, doesn't happen by accident: the food that fails to pass muster in this greengrocer's beauty parade is discarded, despite the fact that it's no less nutritious or tasty.

In the UK, the better-off households produce an average of 5 kilos more waste per week than working-class households. Furthermore, people who live in cities, surrounded by convenience, are much more wasteful than those who live in the country, for whom even a trip to the local shop for a pint of milk requires forward planning. The biggest culprits of all are single, middle-class city-dwellers – Britain's fastest-growing socio-economic group – who also happen to be the biggest consumers of ready meals. A century ago, 80 per cent of household waste was made up of dust and ash. In 2006, packaging accounted for 35 per cent of the weight and 50 per cent of the volume of household waste. Professor Bill Rathje from the University of Arizona is an expert in household refuse. He has highlighted a trend among young urban professionals, which he calls 'the fast lane syndrome'. This group is not only consuming ready meals but is also buying fresh produce in the vain hope that they will make the time to cook it. They kid themselves that ready meals form no more than a standby, but studies show it's the ready meals that get eaten and the fresh meat and vegetables that get thrown away. We produce over 100 million tonnes of rubbish every year in the UK, and household waste is currently increasing at a rate of 3 per cent a year.

In practice, reversing this trend would require little more than going up the supply chain and looking for ways to cut waste and recycle at every step. For our part, we will have to change what we eat and become less wasteful – and if we are throwing away less, we are eating more of what we buy and therefore spending less as well. It really is that straightforward. Unfortunately, advice to 'consume less and spend less' is unlikely to chime with the shareholders of the big grocery retail chains.

We can go much further. Producing food for exclusively local consumption significantly reduces the level of emissions. The high-tech GM agriculture of the future needs to be augmented by local horticulture that educates as well as feeds the population. Reconnecting people with the food supply is essential if we are going to become convinced of its value. A great example of what can be achieved is taking place in the market town of Todmorden in Calderdale, West Yorkshire. A community growers' group has taken over all public spaces and converted them to places of food production. They began with simple herb gardens but are now successfully growing fruit and vegetables in green spaces all over the town: the railway and fire stations, parks and gardens; even in front of the town hall. Every school is involved in the project, which also serves as a major food-based learning programme for the entire community. Even late-night revellers play their part by 'contributing ammonia' to the maize growing outside the police station!

We are in a position where we can still make choices about how our food is produced in the future and what we eat, but they are neither the ones that we might imagine nor those that dominate the mass media. It's not a question of GM vs. Organic, Vegans vs. Omnivores, Farms vs. Forest. It's not necessary for any of the Earth's projected 9 billion population to starve, nor for us to ruin the environment in the process of trying to feed them. A target of 3,000 calories per person per day is an entirely realistic yield, even as farms take steps to protect the environment by reducing reliance on fossil fuels and calling a halt to deforestation. But if we want to continue to enjoy a diet that includes meat and dairy, then we will have to embrace GM and consume a lot less.

In the future, the global unit of currency will not be the dollar, but the calorie. We need to think of food in terms of trading energy, as a matter of fact rather than merely on some conceptual level. So, to highlight the moral and economic challenges of maintaining a diet high in meat and dairy for an increasing proportion of the world's population, consider the energy involved in making a cup of tea, white with one sugar. Boiling the kettle contributes 40 per cent of the energy and 20 per cent comes from growing, processing, packaging and transporting the tea and sugar. The remaining 40 per cent comes from the splash of milk!

Dairy worked well when the Earth's population was less than 100 million people, but it cannot support a global population of 9 billion. Dairy can be considered anachronistic, not a mistake necessarily – at the time it was an excellent solution to the problem of getting more calories off a finite piece of land and then having a high-energy-density food for storage and trade – but it has become hard-wired into our culture: a question of taste, rather than of cost. We have no moral stance on dairy and merely present this as a statement of truth. We should recognise that dairy is a want, not a need; and if we want it to continue, we should accept that we are going to have to pay much more for it and consume much less of it.

The new attitudes to food that will emerge over the next few years will be familiar to those of us who grew up in the 1960s and 70s, when waste was rarely countenanced and little was thrown away. Our relationship with food is going to change, but it can be argued that these changes will be entirely beneficial, resulting in healthier diets all round, which will have positive implications not just for ourselves but for our society as a whole. Obesity is not just a problem for individuals but for the whole nation, accounting for a significant proportion of the healthcare budget – and it has also been identified as one cause of declining economic productivity.

We already know how to make plants that are better suited to the available farmland, but we need to accept that the introduction of GMs with increased yields that require fewer chemicals is a necessity, not a luxury. Again, we can easily go much further. We can match plants to changing environments and address some of the issues associated with climate change by designing crops that are the optimum for a particular environment. As well as GM we have the opportunity to manage the entire ecosystem. In the past, plant-breeding has never looked at the ecosystem holistically, rather it has focused on single traits (e.g. yield per hectare), so crops have lost their connection with the environment. In modern agriculture the soil plays no other role than providing something for the fertiliser to land on. The soil is a precious natural resource just like uranium, oil or coal. There is a limited supply and it needs to treated with care and respect. The people best placed to do that are those who understand what it means to live off the land: the farmers. Our challenge is to develop a global agricultural system that has this local knowledge at its heart, and seeks to build upon it rather than obliterate it. This should not be misconstrued as a desire to move towards monoculture.

We can import some of the successful symbiotic relationships employed by meadow flowers and other wild plants with fungi, such as the trading of sugar for nutrients. Currently, horticultured plants can't do this – they rely on interventions by farmers to do it for them – but we can breed strains that will be able do so (and in this case we wouldn't even need to genetically modify them).

Food is going to get more expensive, but given the amount of food currently produced that is thrown away, it's fair to say that curbing our profligacy will help to offset the increase. For that reason, our diet is going to change: we will eat more vegetables, consume much less meat, fish and dairy, and get more of our protein from grains and fungi. Taking these strong countermeasures to prevent a future food crisis will also alleviate current problems before they get worse. Relying heavily on demand-side economics will not sort the problem out globally. Price-rise signals in the developed world are too weak: people may complain about them but they will ultimately just ignore them, and there is inertia and inefficiency across the whole of our food and agriculture system. Real changes can occur only if we begin to regard malnutrition in geographically remote under-developed nations as a high-priority problem for those of us fortunate enough to be living in the richer countries.

We need to accept that the concept of everlasting

exponential growth is a fallacy. It's not possible to achieve a rate of steady growth in people or physical output in perpetuity. Just 1 per cent growth compounded over 3,000 years would multiply people and possessions by 7 trillion times the original number. There is a ceiling of just 400 years' growth in energy use at a modest 2.3 per cent a year (which is much less than we are actually experiencing at the moment) due to the thermodynamic effect alone, but escalating effects of climate change will have put an end to us long before that is ever reached. How can we possibly have exponential growth on a finite planet? Yet few politicians are willing to deal with that reality, and even fewer economists. The problems we face are huge, but not insurmountable.

At the moment.

The longer we delay in our response to resource shortages, especially in terms of the investment needed for renewable energy and agriculture, the more severe they will become.

15 Project Sunshine

'When a great team loses through complacency, it
will constantly search for new and more intricate
explanations to explain away defeat.'
Pat Riley, NBA player and coach

When we began this book, our aim was to explore how we
might address the four biggest challenges of the 21st century: population growth; food security; energy security; and
climate change. It was a daunting undertaking. These are
scary subjects, and when we are scared, most of us find our
natural inclination is for flight rather than fight. If we can
ignore the problem rather than confront it, then so much
the better.

Certainly at the outset we shared this sense of trepida-
tion, but having wrestled with these subjects for the best part
of a year together, we are now much more optimistic about
the future than we were at the beginning. There is a lot being
done to tackle all of these issues. But we could be doing an
awful lot more. And we need to start now. We shouldn't be
scared of the future: the challenges we face are considerable,
not insurmountable. There are practical solutions available,
but we cannot afford to do nothing. And so, to the four
challenges we would like to add a fifth, arguably the biggest
of all: complacency. The only choice we have is to act sooner
rather than later. That we must change the way we live is simply a matter of fact, it's not really up for discussion: maintaining the status quo is not an option. But given where we are
at the moment – boom and bust, economic strife, crippling
debt and widespread unemployment – there's no reason to
assume that this change need be for the worse. We should be
aiming to make life better for everybody.

The continuing depletion of the Earth's finite natural
resources will necessarily lead to sustainability rather than

growth becoming the key economic driver. In the developed world, there will have to be a reduction in consumption of resources. This can either be a painful realignment due to increased competition from the developing economies driving up prices, or a pro-active readjustment based on a policy of 'reduce, re-use and recycle'. Conversely, consumption in the under-developed world needs to be increased so that the poorest billions can be lifted out of poverty. This will in turn create new markets and opportunities for products and services. None of this will be achieved by relying on market forces exclusively to provide a solution.

While in terms of what needs to be done the science is well understood, the know-how to turn 'ground-breaking research into a commercial proposition' is often lacking. This is simply a question of funding, or rather lack of it (particularly in the case of solar and fusion, where it's woefully inadequate). Delivering short-term shareholder value won't do it. Here, there's a clear role for governments to play, both nationally and globally. Policy should be holistic and international in scope. No one country can solve these problems on its own. Success will undoubtedly require international legislation, but also collaboration in research and development across a range of traditional disciplines.

Stabilising population growth must be the priority. The way to achieve this is through better, and freely available, family-planning advice throughout the developing world (wherein most of the growth is occurring). Most important of all is providing better education for girls and women. These are two eminently achievable targets that would allow us to beat the population bulge before it happens and provide tangible benefits for the whole world.

In the West, we need to start paying our way. That means prices that reflect the true cost of the resources we are taking from the developing world, now and for ever, and not their net-present-value. We are going to end up spending more on less, but our goals should be about quality not quantity. Clearly, in a society that throws away a third of all food there is great scope for improvement. We need to throw away less. Our aim should be to re-use whenever possible and to recycle everything.

While there are many different paths along the road to a solar energy economy, it's clear that this has to be our ultimate destination. The transition from carbon to solar is going to take time, but we cannot afford to delude ourselves any longer that it need not happen at all. The truth is that the sooner we can wean ourselves off fossil fuels, the better for everyone. There are signs that the greatest of all economic drivers, the US military, is starting to take the issue of peak oil very seriously indeed. The Pentagon is one of the most pragmatic institutions, so it should not be imagined that this shift is being motivated by any desire to 'make war greener'. Rather, it's a direct response to the fact that the world's remaining supplies of oil are mostly located in the most politically unstable regions. How ironic that an organisation capable of destroying our world might well end up saving it.

While we continue to burn carbon in the short term, we need to ensure that we are doing so in order to ramp up the alternatives. On balance, and despite all the drawbacks, we should continue to invest in wind, hydro and other renewable forms of energy. They are not the answer in themselves, but the more we can bring online, the less we will have to rely on coal, gas and oil. Renewable energy should be viewed as (a small) part of an integrated energy supply.

For the same reason, we see no alternative but to increase our global nuclear fission capacity at least three-fold over the next 30 years. Nuclear is the only scaleable alternative to fossil fuels that doesn't suffer from intermittency. Nuclear can be made safer, and it needs to be made safer, but the opportunity it gives us to reduce carbon emissions in the short term is too great to ignore. Pound-for-pound, fusion would provide us with much more energy than any other technology currently in the market or in development. And we need to make it so dollar-for-dollar. The fuel itself is certainly abundant, giving us the potential for millions of years of energy. However, we are effectively still in the very early stages of development. Given the technological challenges, even the astronomical sums invested to date are barely enough. If we are serious about fusion we probably need an investment in the order of $1 trillion to discover whether it can really work. Even that commitment would come without guarantees: we might well end up concluding that we simply need to forget about it altogether.

Research into all aspects of solar energy – photovoltaics, photosynthesis, liquid fuels – needs to be globally coordinated. Delivering solar power at scale in the timeframe required is possible, but it needs massive investment today. Solar must be viewed as a must-have, not a nice-to-have. It should no longer be viewed as an 'alternative' form of energy, but as the only form of energy.

The days of throwing away a third of the food we produce are coming to an end. We have enjoyed three decades of cheap prices, but the cost of the average weekly shop is forecast to double in real terms over the next 20 years. A worldwide collective agricultural policy, in practice based on high-tech farming, GM crops and no-till, would allow us to sustain a global population of 9 billion. This should be augmented by local food-growing schemes, on public spaces in cities and towns, to reconnect the population with what food means. We will not have to become vegans, but we will have to change what we eat. Since 1980, bread consumption in Europe and the US has been in decline as consumption of chicken, beef and pork has increased. Bread is no longer the staff of life; it has become something we wrap food in. The demand for meat is increasing at an unsustainable rate, which alone will ensure that average consumption is reduced. Meat has to be part of the global food web, but on a reduced scale. Using grain for animal feed is an inefficient luxury we can no longer afford. Meat and dairy will still appear on our plates, but probably not every day, although it's unlikely that we are going to be able to afford organic dairy at any scale at all.

Again, none of this is necessarily bad news. The health and vitality of the British civilian population peaked during the years of rationing during and after the Second World War. There is no need to return to the austerity of that era, but reducing the levels of meat and dairy in our diets will have many positive consequences.

Food production itself needs to be decoupled from oil. GM can reduce the reliance on petrochemicals for herbicide and pesticide, and when combined with the lessons learned from organic agriculture can also reduce the need for artificially produced fertiliser. Using local produce is of paramount importance. We should respect the seasons – there ought to be no more strawberries in winter – and the only food that should fly is game and poultry.

Science can help us to fuel and feed the world, but the easiest way to address all the challenges is to reduce consumption and use the resources we have much more efficiently. There are huge savings that can be made easily. We use seven times as much water as we did in 1900, produce sixteen times the number of manufactured goods and transport them by aeroplane over 1.5 trillion passenger air miles each year, yet the vast majority of these goods – around 80 per cent – are used just once and then discarded. Rather than increasing at the same rate as the global population, consumption is not only outpacing it but speeding up as it does so. China alone is currently adding the equivalent of two large power stations every week to feed its economic growth. A culture of reducing consumption needs to become ingrained. We should consider waste and inefficiency to be morally abhorrent. Our goal should be nothing less than recycling every atom.

Until 350 years ago, mankind was living like any other animal, beholden to the solar cycle. That we will return to that situation in the near future is inevitable. And when we reach that point, the only question will be whether we took the right decisions to fuel and feed a world of 9 billion people.

References, Sources and Further Reading

Preface: Tick Tock
'Time to wake up', *GMO Quarterly News Letter*, April 2011
The Day Before Yesterday: Five Million Years of Human History, Colin Tudge (Jonathan Cape, 1995), ISBN: 978-0712661737
People and the Planet, The Royal Society Science Policy Centre Report, 2012, ISBN: 978-0854039555

Chapter 1: Seven Billion and Counting
'World's "seven billionth baby" is born', *The Guardian*, 31 October 2011
UFPA, Day of 7 Billion Actions: www.7billionactions.org
World Population Prospects, United Nations Department of Economic and Social Affairs, 2010 revision
People and the Planet, The Royal Society Science Policy Centre Report, 2012, ISBN: 978-0854039555
State of World Population 2012: By Choice, Not by Chance: Family Planning, Human Rights and Development, UNFPA, 2012, ISBN: 978-1618000095
www.worldometers.info/world-population/
An Inconvenient Truth: The Planetary Emergency of Global Warming and What We Can Do About It, Albert J. Gore (Bloomsbury, 2006), ISBN: 978-0-74758-906-8
Population: One Planet, Too Many People, UK Institution of Mechanical Engineers, January 2011; www.imeche.org/Libraries/2011_Press_Releases/Population_report.sflb.ashx
The Road, Cormac McCarthy (Picador, 3rd edn, 2009), ISBN: 978-0330468466

Chapter 2: Weathering a Perfect Storm
Food, Energy, Water and the Climate: a Perfect Storm of Global Events?, John Beddington, UK government Office for Science, 2009; www.bis.gov.uk/assets/goscience/docs/p/perfect-storm-paper.pdf
OECD National Accounts at a Glance 2011, OECD Publishing, 2011, ISBN: 978-9-26412-498-1
The State of Food Insecurity in the World, UNFAO, 2012, ISBN: 978-92-5-107316-2
Oil Market Report, International Energy Agency, February 2012

Chapter 3: Star Power
Cosmos, Carl Sagan (Abacus, 1983), ISBN: 978-0349107035
The Universe in a Nutshell, Stephen Hawking (Bantam Press, 2001), ISBN: 978-0593048153

'Dark Matter WIMPS Out', *New Scientist*, 28 July 2012, p. 5

Why Does E=mc²?, Brian Cox and Jeff Forshaw (Da Capo, 2010), ISBN: 978-0306819117

Revolutions That Made the Earth, Tim Lenton and Andrew Watson (Oxford University Press, 2011), ISBN: 978-0-19-958704-9

Origins of Life, Freeman Dyson (Cambridge University Press, 2nd edn, 1999), ISBN: 978-0521626682

Dark Matter, Dark Energy, Dark Gravity, Stephen Perrenod (FeedBrewer, 2011), ASIN: B004OL2N92

Afterglow of Creation: Decoding the Message from the Beginning of Time, Marcus Chown (Faber & Faber, 2010), ISBN: 978-0571250592.

The Grand Design, Stephen Hawking and Leonard Mlodinow (Bantam, 2011), ISBN: 978-0553819229

A Brief History of Time: From Big Bang to Black Holes, Stephen Hawking (Bantam, 2011), ISBN: 978-0857501004

Teaching About Evolution and the Nature of Science, Working Group on Teaching Evolution (National Academy Press, 1998), ISBN: 978-0309063647

The Origin of Life, Paul Davies (Penguin, 2003), ISBN: 978-0141013022

Almost Like A Whale: The Origin of Species Updated, Steve Jones (Black Swan, 2000), ISBN: 978-0552999588

The Goldilocks Enigma: Why is the Universe Just Right for Life?, Paul Davies (Penguin, 2007), ISBN: 978-0141023267

The Ancestor's Tale: A Pilgrimage to the Dawn of Life, Richard Dawkins (Phoenix, new edn, 2005), ISBN: 978-0753819968

Coal: A Human History, Barbara Freese (Arrow, 2006), ISBN: 978-0099478843

Party's Over: Oil, War and the Fate of Industrial Societies, Richard Heinberg (Clairview Books, 2005), ISBN: 978-1905570003

Why Your World is About to Get a Whole Lot Smaller: Oil and the End of Globalisation, Jeff Rubin (Virgin Books, 2010), ISBN: 978-0753519639

Chapter 4: Taking Control

Revolutions That Made the Earth, Tim Lenton and Andrew Watson (Oxford University Press, 2011), ISBN: 978-0199587049

The Origin of Our Species, Chris Stringer (Penguin, 2012), ISBN: 978-0141037202

The Incredible Human Journey, Alice Roberts (Bloomsbury, 2010), ISBN: 978-1408802885

Catching Fire: How Cooking Made Us Human, Richard Wrangham (Profile Books, 2009), ISBN: 978-1846682858

The Origin of Humankind, Richard Leakey (Weidenfeld & Nicolson, 1994), ISBN: 978-0297815037

River Out of Eden, Richard Dawkins (Basic Books, 1995), ISBN: 978-0-465-06990-3

Genetics: From Genes to Genomes, Leland Hartwell (McGraw-Hill, 2010), ISBN: 978-0071221924

After the Ice: A Global Human History, 20,000–5000 BC, Steve Mithin (Phoenix, 2004), ISBN: 978-0753813928

First Farmers: The Origins of Agricultural Societies, Peter Bellwood (Wiley-Blackwell, 2004), ISBN: 978-0631205661

The Humans Who Went Extinct: Why Neanderthals Died Out and We Survived, Clive Finlayson (Oxford University Press, 2010), ISBN: 978-0199239191

The Dynamics of Neolithisation in Europe, A. Hadjikoumis, E. Robinson and S. Viner (eds) (Oxbow Books, 2011), ISBN: 978-1842179994

Y: The Descent of Men, Steve Jones (Abacus, 2003), ISBN: 978-0349113890

Chapter 5: States of Emergency

'The Neolithic Revolution: an Ecological Perspective', John C. Barrett, in *The Dynamics of Neolithisation in Europe*, A. Hadjikoumis, E. Robinson and S. Viner (eds) (Oxbow Books, 2011), ISBN: 978-1842179994

Sumer and the Sumerians, Harriet Crawford (Cambridge University Press, 2004), ISBN: 978-0521533386

The Maya (Ancient Peoples and Places), Michael D. Coe (Thames & Hudson, 2011), ISBN: 978-0500289020

The Fall of the Roman Empire: A New History, Peter Heather (Pan, 2006), ISBN: 978-0330491365

Empires and Barbarians: Migration, Development and the Birth of Europe, Peter Heather (Pan, 2010), ISBN 978-0330492553

Hadrian's Empire: When Rome Ruled the World, Danny Danziger and Nicholas Purcell (Hodder & Stoughton, 2005), ISBN: 978-0340833605

How Rome Fell: Death of a Super Power, Adrian Goldsworthy (Yale University Press, 2010), ISBN: 978-0300164268

'The History of Rome', podcast by Mike Duncan; www.thehistoryof rome.typepad.com

'Energy crisis? We've been here before', Colin McInnes, University of Strathclyde, July 2010, www.spiked-online.com

The Time Traveller's Guide to Elizabethan England, Ian Mortimer (Bodley Head, 2012), ISBN: 978-1847921147

A Brief History of Life in the Middle Ages, Martyn Whittock (Robinson, 2009), ISBN: 978-1845296858

The Black Death: The Intimate Story of a Village in Crisis 1345–50: An Intimate History, John Hatcher (Phoenix, 2009), ISBN: 978-0753823071

The Black Death, Philip Ziegler (Sutton Publishing, 2003), ISBN: 978-0750932028

The British Industrial Revolution in Global Perspective, Robert C. Allen (Cambridge University Press, 2009), ISBN: 978-0521687850

Chapter 6: The Solar Deficit

The Age of Revolution: 1789–1848, Eric Hobsbawm (Abacus 1988),
ISBN: 978-0349104843

The Age of Capital: 1848–1875, Eric Hobsbawm (Abacus 1988), ISBN:
978-0349104805

The Age of Empire: 1875–1914, Eric Hobsbawm (Abacus, 1989), ISBN:
978-0349105987

The Birth of the Modern World, 1780–1914, C.A. Bayly
(Wiley-Blackwell, 2004), ISBN: 978-0631236160

An Essay on the Principle of Population, Thomas Malthus (Oxford
Paperbacks, 1999), ISBN 978-0192837479

Malthus Past and Present, J. Dupaquier (ed.) (Academic Press, 1983),
ISBN: 978-0122246708

William Godwin: A Short Biography, Leslie Stephen (Shamrock Eden,
2011), ASIN: B0069XER9M

Eruptions That Shook the World, Clive Oppenheimer (Cambridge
University Press, 2011), ISBN: 978-0521641128

'Imagination and Innovation of an Industrial Pioneer: The First
Abraham Darby', Nancy Cox, *Industrial Archaeology Review*, 1990,
12 (2), pp. 127–44

'Darby, Abraham (1678–1717)', Nancy Cox, *Oxford Dictionary of
National Biography* (Oxford University Press, 2004), online edn, Jan
2008; www.oxforddnb.com/view/article/7137, accessed 2 January
2013

In Search of the Dark Ages, Michael Wood (BBC, 2006), ISBN:
978-0563522768

In Search of the First Civilizations, Michael Wood (BBC, 2005), ISBN:
978-0563522669

*Making a Living in the Middle Ages: The People of Britain, 850–
1520*, Christopher Dyer (Yale University Press, 2009), ISBN:
978-0300101911

The British Industrial Revolution in Global Perspective, Robert C. Allen
(Cambridge University Press, 2009), ISBN: 978-0521687850

Iron and Steel, William F. Hosford (Cambridge University Press,
2012), ISBN: 978-1107017986

Victorian London: The Life of a City 1840–1870, Liza Pickard (Phoenix,
2006), ISBN: 978-0753820902

*The Wizard of Menlo Park: How Thomas Alva Edison Invented the
Modern World*, Randall E. Stross (Three Rivers Press, 2007), ISBN
978-1400047635

Wizard: Life and Times of Nikola Tesla, Marc J. Seifer (Citadel Press,
1997), ISBN: 978-1559723299

AC/DC: The Savage Tale of the First Standards War, Tom McNicol
(Jossey Bass, 2006), ISBN: 978-0787982676

Chapter 7: Food, Glorious Food

The Time Traveller's Guide to Elizabethan England, Ian Mortimer (Bodley Head, 2012), ISBN: 978-1847921147

A Brief History of Life in the Middle Ages, Martyn Whittock (Robinson, 2009), ISBN: 978-1845296858

A History of World Agriculture: From the Neolithic Age to the Current Crisis, Marcel Mazoyer and Laurence Roudart (Routledge, 2006), ISBN: 978-1844073993

Essential Soil Science: A Clear and Concise Introduction to Soil Science, Mark Ashman and Geeta Puri (Wiley-Blackwell, 2008), ISBN: 978-0632048854

Darwinian Agriculture: How Understanding Evolution Can Improve Agriculture, R. Ford Dennison (Princeton University Press, 2012), ISBN: 978-0691139500

Agriculture and Food in Crisis: Conflict, Resistance, and Renewal, Fred Magdoff and Brian Tokar (Monthly Review Press, 2010), ISBN: 978-1583672266

The Nitrate King: A Biography of 'Colonel' John Thomas North, William Edmundson (Palgrave Macmillan, 2011), ISBN: 978-0230112803

Encyclopedia of Agricultural Science: Volumes 1–4, Charles J. Arntzen and Ellen M. Ritter (eds) (Academic Press, 1994), ISBN: 978-012226670

Enriching the Earth: Fritz Haber, Carl Bosch, and the Transformation of World Food Production, Vaclav Smil (MIT Press, 2004), ISBN: 978-0262693134

Between Genius and Genocide: The Tragedy of Fritz Haber, Father of Chemical Warfare, Dan Charles (Jonathan Cape, 2005), ISBN: 978-0224064446

Geopolitics and the Green Revolution: Wheat, Genes, and the Cold War, John H. Perkins, Oxford University Press USA, 1998), ISBN: 978-0195110135

Our Daily Bread: The Essential Norman Borlaug, Noel Vietmeyer (Bracing Books, 2011), ISBN: 978-0578095554

'Feeding a World of 10 Billion People: The Tva/Ifdc Legacy', Norman Borlaug, *International Fertilizer Development*, 2003, ISBN: 978-0880901444; ifdc.org/getdoc/262b2fea-2f9e-4598-b29e-25358eb06511/LS-3--Feeding_a_World_of_10_Billion_People.aspx

Chapter 8: When the Explosion Stops

State of World Population 2012: By Choice, Not by Chance: Family Planning, Human Rights and Development, UNFPA, 2012, ISBN: 978-1618000095

United States Census Bureau, www.census.gov

Everything Now, Steve McKevitt (Route Publishing, 2012), ISBN: 978-1901927511

Scarcity and Frontiers: How Economies Have Developed Through Natural Resource Exploitation, Edward B. Barbier (Cambridge University Press 2010), ISBN: 978-0521701655

Pay Check: Are Top Earners Really Worth It?, David Bolchover (Coptic, updated edn, 2012), ASIN: B006YVTTOM

The Living Dead: Switched Off, Zoned Out – The Shocking Truth About Office Life, David Bolchover (Capstone, 2005), ISBN: 978-1841126562

City Slackers, Steve McKevitt (Cyan, 2006), ISBN: 978-1904879725

en.wikipedia.org/wiki/World_population (downloaded 11 December 2012)

'A Theory of Human Motivation', A.H. Maslow, *Psychological Review*, 1943, 50, pp. 370–96

People and the Planet, The Royal Society Science Policy Centre Report, 2012, ISBN: 978-0854039555

'Human Population Grows Up', J. Cohen, *Scientific American*, 2005, 293, pp. 48–55

'The Impact of Population Growth on Tomorrow's World', M. Potts, A. Pebley and J. Speidel, *Philosophical Transactions of the Royal Society B*, 2009, 364, pp. 2975–76

'Economic and Social Implications of the Demographic Transition', D. Reher, *Population and Development Review*, 2011, 37, (S1), pp. 11–33

'The Unfolding Story of the Second Demographic Transition', R. Lesthaeghe, *Population and Development Review*, 2010, 36, 2, pp. 211–51

'Women's Work and Economic Development', K. Mammen and C. Paxson, *The Journal of Economic Perspectives*, 2000, 14 (4), pp. 141–64

'A Safe Operating Space for Humanity', J. Rockström, W. Steffen, K. Noone, A. Persson, F.S. Chapin, E.F. Lambin, T.M. Lenton, M. Scheffer, C. Folke, H.J. Shellnhuber, B. Nykvist, C.A. de Wit, T. Hughes, S. van der Leeuw, H. Rodhe, S. Sorlin, P.K. Snyder, R. Costanza, U. Svedin, M. Falenmark, L. Karlberg, R.W. Correll, V.J. Fabry, J. Hansen, B. Walker, D. Liverman, K. Richardson, P. Crutzen and J.A. Foley, *Nature*, 2009, 461, pp. 472–5

Mismeasuring Our Lives: Why GDP Doesn't Add Up, Joseph Stiglitz, Amartya Sen and Jean-Paul Fitoussi (New Press, 2010), ISBN: 978-1595585196

'Population Priorities: The Challenge of Continued Rapid Population Growth', A. Turner, *Philosophical Transactions of the Royal Society B*, 2009, 364, pp. 2977–84

Levels and Trends in Child Mortality Report 2011, UNICEF on behalf of the UN Inter-agency Group for Child Mortality Estimation; www. who.int/maternal_child_adolescent/documents/20110915_unicef_childmortality/en/index.html

Population, Reproductive Health and the Millennium Development Goals, UNDP, 2005; www.unmillenniumproject.org/documents/ SRHbooklet080105.pdf

'Harvesting the Biosphere: The Human Impact', Vaclav Smil,
  Population and Development Review, 2011, 37 (4), pp. 613–36
'How Many Billions to Go?', Vaclav Smil, *Nature*, 1999, 401, p. 429

Chapter 9: What's So Good About Oil

'Powering the Planet', N.S. Lewis, *MRS Bulletin*, 2007, 32, pp. 808–20
'Transocean Ltd Provides Deepwater Horizon Update' (press release),
  Transocean Ltd, 26 April 2010 (retrieved 21 May 2010)
'Deepwater Horizon: A Timeline of Events', Offshore-Technology
  (Net Resources International), 7 May 2010 (retrieved 21 May 2010)
www.bbc.co.uk/news/special_reports/oil_disaster/
www.guardian.co.uk/environment/bp-oil-spill
'BP Disaster: Worst Oil Spill in US History Turns Seas Into a Dead
  Zone', *Daily Telegraph*, 29 May 2010
'The Real Cost of Keeping Warm', Tim Harford, *Financial Times*,
  12 November 2011
Energy Cost Impacts on American Families, 2001–2012, American
  Coalition for Clean Coal Energy, February 2012
Fuel Poverty: Changing the Framework for Measurement, UK consult-
  ation document, Department of Energy and Climate Change, 2012,
  ISBN: 978-0101844024
'UK Household Spending Crippled by Energy Bills', Shane Croucher,
  International Business Times, 4 December 2012
James Joule: A Biography, D.S.L. Cardwell (Manchester University
  Press, 1989), ISBN: 978-0719030253
Sustainable Energy – Without the Hot Air, David JC MacKay (UIT
  Cambridge, 2008), ISBN: 978-0954452933
CRC Handbook of Chemistry and Physics, David R. Lide (ed.) (CRC
  Press, 86th revised edn, 2005), ISBN: 978-0849304866
Physics for Future Presidents: The Science Behind the Headlines, Richard
  A. Muller (W.W. Norton, 2008), ISBN: 978-0393066272
*Physics and Technology for Future Presidents: An Introduction to the
  Essential Physics Every World Leader Needs to Know*, Richard A.
  Muller (Princeton University Press, 2010), ISBN: 978-0691135045
The Quest: Energy, Security, and the Remaking of the Modern World,
  Daniel Yerg (Penguin, 2012), ISBN: 978-0143121947
Oil: A Beginner's Guide, Vaclav Smil (Oneworld, 2008), ISBN:
  978-1851685714
*Prime Movers of Globalization: The History and Impact of Diesel
  Engines and Gas Turbines*, Vaclav Smil (MIT Press, 2010), ISBN:
  978-0-262-01443-4
*Transforming the Twentieth Century: Technical Innovations and Their
  Consequences*, Vaclav Smil (Oxford University Press, 2006), ISBN:
  978-0195168754
Energy at the Crossroads: Global Perspectives and Uncertainties, Vaclav
  Smil (MIT Press, 2006), ISBN: 978-0262693240

Energy in Nature and Society: General Energetics of Complex Systems,
    Vaclav Smil (MIT Press, 2007), ISBN: 978-0262693561
Energy Transitions: History, Requirements, Prospects, Vaclav Smil
    (Praeger, 2010), ISBN: 9780313381775
'Peak Oil: A Catastrophist Cult and Complex Realities', Vaclav Smil,
    World Watch, 2006, 19, pp. 22–4
Saving the Planet Through Pesticides and Plastics, Dennis T. Avery
    (Hudson Institute, 2nd edn, 2000), ISBN: 978-1558130692
Physical Chemistry, Peter Atkins and Julio de Paula (Oxford University
    Press, 2009), ISBN: 0199543372
'Climate Policy: Oil's Tipping Point Has Passed', J. Murray and D.
    King, *Nature*, 2012, 481, pp. 433–5
'Plows, Plagues, and Petroleum: How Humans Took Control of
    Climate', Vaclav Smil, *International History Review*, 2006, 28,
    pp. 931–2
www.formula1.com/inside_f1/understanding_the_sport/8763.html
'A Plan to Power 100 Percent of the Planet with Renewables', Mark
    Z. Jacobson and Mark A. Delucchi, *Scientific American*, November
    2009, pp. 58–65
*Contribution of Working Groups I, II and III to the Fourth Assessment
    Report of the Intergovernmental Panel on Climate Change*,
    R.K. Pachauri and A. Reisinger (eds) (IPCC, 2008), ISBN:
    978-9291691227

Chapter 10: Going Nuclear

hansard.millbanksystems.com/commons/1950/nov/20/advertising-
    general-policy
www.tonybenn.com/nucl.html
blueandgreentomorrow.com/2012/2/20/questions-of-efficiency/
dae.nic.in; government of India, Department of Atomic Energy
www.iaea.org; International Atomic Energy Agency
www.euronuclear.org; European Nuclear Society
www.aecl.ca; Atomic Energy of Canada Ltd
www.nrc.gov; United States Nuclear Regulation Committee
Risk Management: The Nuclear Liabilities of British Energy plc, National
    Audit Office, 6 February 2004 (retrieved 25 August 2006)
'The dream that failed', *The Economist*, 10 March 2012
Radioactive Curative Devices and Spas, Paul W. Frame (Oak Ridge
    Associated Universities, 1989); www.orau.org/ptp/articlesstories/
    quackstory.htm
Madame Curie: A Biography, Eve Curie Labouisse (Da Capo, new edn,
    2001), ISBN: 978-0306810381
news.nationalgeographic.co.uk/news/energy/2011/04/
    pictures/110426-chernobyl-25th-anniversary-wildlife/#/chernobyl-
    wildlife-chickens-cat_34167_600x450.jpg

'25 Years On, Chernobyl Lakes Thriving Despite Fallout': www.
newscientist.com/article/dn20418-25-years-on-chernobyl-lakes-
thriving-despite-fallout.html
'Wildlife and Chernobyl: The Scientific Evidence For Minimal
Impacts', Robert J. Baker, Jeffrey K. Wickliffe, *Bulletin of Atomic
Scientists*, 14 April 2011
'Thorium Reactors Could Rescue Nuclear Power – Special Report',
David Shinga, *New Scientist*, 23 March 2011
'India's Thorium-Based Nuclear Dream Inches Closer', Hal Hodson,
New Scientist, 9 November 2012
'Nuclear Alchemy: Thorium Promises Power From Waste', *New
Scientist*, 30 May 2012
'The History of Nuclear Power', R.J. Duffy, in *Encyclopedia of Energy*,
Cutler J. Cleveland (ed.) (Elsevier, 2004), pp. 395–408, ISBN:
978-0-12-176480-7
*Nuclear Energy: An Introduction to the Concepts, Systems,
and Applications of Nuclear Processes*, Raymond L. Murray
(Butterworth-Heinemann, 2008), ISBN: 978-0123705471
Power to Save the World: The Truth about Nuclear Energy, Gwyneth
Cravens (Vintage Books USA, 2008), ISBN: 978-0307385871
*Atomic Awakening: A New Look at the History and Future of
Nuclear Power*, James Mahaffey (Pegasus Books, 2007), ISBN:
978-1605981277
Uranium, Tom Zoellner (Penguin USA, 2010), ISBN:
978-0143116721
The Nuclear Fuel Cycle: From Ore to Waste, P D. Wilson (ed.) (Oxford
University Press, 1996), ISBN: 978-0198565406

Chapter 11: Tilting at Windmills

The Secret Life of the National Grid, Gaby Hornsby, BBC TV series, 2010
www.theiet.org; The Institute of Engineering and Technology
Archive: 'British Electrical Power Convention: "Electricity in the
Modern Home"', Dame Caroline Haslett, 1954, UK0108 NAEST
033/2/17
'National Trust Comes Out Against 'Public Menace of Wind Farms',
Louise Gray, *Daily Telegraph*, 12 February 2012
Energy Cost Impacts on American Families, 2001–2011, Eugene M.
Trisko, study commissioned by American Coalition for Clean Coal
Energy, January 2011
'Embracing the Wind: Denmark's Recipe for a Model Democracy',
Manfred Ertel and Gerald Trauffetter, *Der Spiegel International*, 24,
August 2012
Biofuels: Ethical Issues, Nuffield Council on Bioethics (Nuffield Press,
2011); www.nuffieldbioethics.org/sites/default/files/Biofuels_ethi-
cal_issues_FULL%20REPORT_0.pdf
Sustainable Energy – Without the Hot Air, David J.C. MacKay (UIT
Cambridge, 2008), ISBN: 978-0954452933

'Why Wind Power Works for Denmark', H. Sharman, *Proceedings of the Institution of Civil Engineers (ICE), Civil Engineering*, 2005, 158, pp. 66–72

The Economics of Climate Change: The Stern Review, Nicholas Stern (Cambridge University Press, 2007), ISBN: 978-0521700801

Hoover Dam: An American Adventure, Joseph E. Stevens (University of Oklahoma Press, 1990), ISBN: 978-0806122830

Geothermal Energy Systems: Exploration, Development, and Utilization, Ernst Huenges and Patrick Ledru (eds) (Wiley-VCH, 2010), ISBN: 978-3527408313

Energy for a Sustainable World: From the Oil Age to a Sun-Powered Future, Vincenzo Balzani and Nicola Armaroli (Wiley-VCH, 2010), ISBN 978-3-527-32540-5

Renewable Energy, Godfrey Boyle (Oxford University Press, 2004), ISBN: 978-0199261789

Building the Three Gorges Dam, L. Patricia Kite (Raintree, 2011), ISBN: 978-1406220230

'Large-scale Impacts of Hydroelectric Development', D.M. Rosenberg, F. Berkes, R.A. Bodaly, R.E. Hecky, C.A. Kelly, J.W.M. Rudd, *Environmental Reviews*, 1997, 5 (1), pp. 27–54

'A Skeptic Looks at Alternative Energy', Vaclav Smil, *IEEE Spectrum*, July 2012, pp. 46–52

Energy Transitions: History, Requirements, Prospects, Vaclav Smil (Praeger, 2010), ISBN: 9780313381775

www.vaclavsmil.com/wp-content/uploads/docs/smil-article-power-density-primer.pdf

Chapter 12: Shine

Cosmos, Carl Sagan (Abacus, new edn, 1983), ISBN: 0349107033

Archimedes' Death Ray Revisited, *Mythbusters*, Season 3, Episode 4, Richard Dowleam, Beyond Productions, 2007

Solar, Ian McEwan (Vintage, 2011), ISBN: 978-0099549024

Chasing the Sun, Richard Cohen (Simon & Schuster, 2011), ISBN: 978-1416526124

Energy at the Crossroads, Vaclav Smil (MIT Press, 2005), ISBN: 978-0262693240

Physics and Technology for Future Presidents, Richard A. Muller (Princeton University Press, 2010), ISBN: 978-0691135045

Handbook of Photovoltaic Science and Engineering, Antonio Luque and Steven Hegedus (eds) (Wiley, 2011), ISBN: 978-0470721698

Third Generation Photovoltaics, Martin Green (Springer, 2003), ISBN 978-3540265627

Sustainable Energy – Without the Hot Air, David J.C. MacKay (UIT Cambridge, 2008), ISBN: 978-0954452933

'Toward Cost-Effective Solar Energy Use', N.S. Lewis, *Science*, 2007, 315, pp. 798–801

'Materials Interface Engineering for Solution-processed Photovoltaics',
M. Grätzel, R.A.J. Janssen, D.B. Mitzi, E.H. Sargent, *Nature*, 2012,
488, pp. 304–12
'Dye-sensitized Solar Cells: a Brief Overview', M.K. Nazeeruddin, E.
Baranoff, M. Grätzel, *Solar Energy*, 2011, 85, pp. 1172–8
'Polymer: Fullerene Bulk Heterojunction Solar Cells', J. Nelson,
Materials Today, 2011, 14, pp. 462–70
'Detailed Balance Limit of Efficiency of P-n Junction Solar Cells',
W. Shockley, H.J. Queisser, *Journal of Applied Physics*, 1061, 32,
pp. 510–19
'Polymer Photovoltaic Cells: Enhanced Efficiencies via a Network
of Internal Donor-acceptor Heterojunctions', G. Yu, J. Gao, J.C.
Hummelen, F. Wudl, A.J. Heeger, *Science*, 1995, 270, pp. 1789–91
'Practical Roadmap and Limits to Nanostructured Photovoltaics',
Richard R. Lunt, Timothy P. Osedach, Patrick R. Brown, Jill
A. Rowehl, Vladimir Bulovi, *Advanced Materials*, 2011, 23,
pp. 5712–27
'A Phase Diagram of the P3HT:PCBM Organic Photovoltaic System:
Implications for Device Processing and Performance', Paul E.
Hopkinson, Paul A. Staniec, Andrew J. Pearson, Alan D.F. Dunbar,
Tao Wang, Anthony J. Ryan, Richard A.L. Jones, David G. Lidzey,
Athene M. Donald, *Macromolecules*, 2011, 44, pp. 2908–17
'Solar Cells: Correlating Structure with Function in Thermally
Annealed PCDTBT:PC70BM Photovoltaic Blends', Tao Wang,
Andrew J. Pearson, Alan D.F. Dunbar, Paul A. Staniec, Darren C.
Watters, Hunan Yi, Anthony J. Ryan, Richard A.L. Jones, Ahmed
Iraqi, David G. Lidzey, *Advanced Functional Materials*, 2012, 22,
pp. 1399–1408
'High-performance Si Microwire Photovoltaics', M.D. Kelzenberg,
D.B. Turner-Evans, M.C. Putnam, S.W. Boettcher, R.M. Briggs, J.Y.
Baek, N.S. Lewis, H.A. Atwater, *Energy and Environmental Science*,
2011, 4, pp. 866–71
www.vaclavsmil.com/wp-content/uploads/docs/smil-article-power-
density-primer.pdf

Chapter 13: Whatever Gets You Through the Night

The World Factbook 2013, Central Intelligence Agency (Skyhorse
Publishing, 2012), ISBN:1616088230
www.rsc.org/ScienceAndTechnology/Policy/Documents/solar-fuels.asp
www.guardian.co.uk/environment/blog/2012/may/11/scientists-
cost-artificial-leaf
www.guardian.co.uk/environment/2009/aug/11/
artificial-leaf-energy?intcmp=239
www.torresolenergy.com/TORRESOL/gemasolar-plant/en
www.electrochem.org/dl/interface/fal/fal10/fal10_p049-053.pdf

'Batteries for Large-Scale Stationary Electrical Energy Storage', Daniel
H. Doughty, Paul C. Butler, Abbas A. Akhil, Nancy H. Clark, John
D. Boyes, *The Electrochemical Society Interface*, Fall 2010, pp. 49–53
www.bbc.co.uk/news/science-environment-20420557
'Magnesium-Antimony Liquid Metal Battery for Stationary Energy
Storage', David J. Bradwell, Hojong Kim, Aislinn H.C. Sirk, et al.,
Journal of the American Chemical Society, 2012, 134, pp. 1895–7
'Powering the Planet: Chemical Challenges in Solar Energy
Utilization', N.S. Lewis and D.G. Nocera, *Proceedings of the
National Academy of Sciences of the United States of America*, 2006,
103, pp. 15729–35
'Toward Cost-effective Solar Energy Use', N.S. Lewis, *Science*, 2007,
315, pp. 798–801
'Solar Energy Conversion', G.W. Crabtree and N.S. Lewis, *Physics
Today*, 2007, 60, pp. 37–42
'Solar Water Splitting Cells', M.G. Walter, E.L. Warren, J.R. McKone,
S.W. Boettcher, Q.X. Mi, E.A. Santori, N.S. Lewis, *Chemical
Reviews*, 2010, 110, pp. 6446–73
'Modeling, Simulation, and Design Criteria for Photoelectrochemical
Water-Splitting Systems', S. Haussener, C. Xiang, J.M. Spurgeon, S.
Ardo, N.S. Lewis, A.Z. Weber, *Energy and Environmental Science*,
2012, 5, pp. 9922–35
'Recent Advances in Hybrid Photocatalysts for Solar Fuel Production',
Phong D. Tran, Lydia H. Wong, James Barber, Joachim S.C. Loo,
Energy and Environmental Science, 2012, 5, pp. 5902–18
www.rsc.org/chemistryworld/2012/11/lithium-air-batteries
'A Reversible and Higher-Rate LiO$_2$ Battery', Zhangquan Peng, Stefan
A. Freunberger, Yuhui Chen, Peter G. Bruce, *Science*, 2012, 337,
pp. 563–6
'Lithium–Air Battery: Promise and Challenges', G. Girishkumar, B.
McCloskey, A.C. Luntz, S. Swanson, W. Wilcke, *Journal of Physical
Chemistry Letters*, 2010, 1, pp. 2193–2203
spectrum.ieee.org/energy/renewables/lithiumair-batteries-get-a-
recharge
Beyond Oil and Gas: The Methanol Economy, G.A. Olah, A.
Goeppert, G.K.S Prakash, G.K. Surya (Wiley-VCH, 2009), ISBN:
978-3-527-32422-4
'Turning Carbon Dioxide Into Fuel', Z. Jiang, T. Xiao, V.L. Kuznetsov,
P.P. Edwards, *Philosophical Transactions of the Royal Society A*, 2010,
368, pp. 3343–64
'Air + water = gasoline? Not quite …', Brian Dodson, October 2012:
www.gizmag.com/air-fuel-synthesis-gasoline-from-air/24739/
www.theengineer.co.uk/sectors/energy-and-environment/news/
cryogenic-energy-storage-plant-could-provide-valuable-back-up/
1007539.article
www.shell.com/home/content/future_energy/scenarios/2050/

www.shell.com/static/future_energy/downloads/shell_scenarios/
shell_energy_scenarios_2050.pdf

www.bp.com/liveassets/bp_internet/globalbp/globalbp_uk_english/
reports_and_publications/statistical_energy_review_2008/STAGING/
local_assets/2010_downloads/2030_energy_outlook_booklet.pdf

'Be Persuasive. Be Brave. Be Arrested (If Necessary)', Jeremy
Grantham, *Nature*, 2012, 491, p. 303

Chapter 14: Feast or Famine?

The Emerald Planet: How Plants Changed Earth's History, David
Beerling (Oxford University Press, 2008), ISBN: 978-0199548149

'GM Food: We Can No Longer Afford to Ignore its Advantages',
Robin McKie, *The Guardian*, 13 October 2012

blogs.scientificamerican.com/science-sushi/2011/07/18/
mythbusting-101-organic-farming-conventional-agriculture/

'Meat in a Low-Carbon World', news.bbc.co.uk/1/hi/sci/tech/
7389678.stm?wwparam=1349466442

Facts and Figures on Obesity, UK Department of Health, 30 April 2012;
www.dh.gov.uk

The Evolution of the Supermarket Industry From A&P to Wal-Mart,
Paul B. Ellickson (University of Rochester, April 2011);
paulellickson.com/SMEvolution.pdf

The State of Food Insecurity in the World 2012, Food and
Agriculture Organization of the United Nations, 2012, ISBN:
978-92-5-107316-2

'UN Warns of Looming Worldwide Food Crisis in 2013', John Vidal,
The Observer, 13 October 2012

Rats, Mice and People: Rodent Biology and Management, Grant R.
Singleton, Lyn A. Hinds, Charles J. Krebs, Dave M. Spratt (eds)
(Australian Centre for International Agricultural Research, 2003),
ISBN: 978-1-86320-357-9

*Crop Prospects and Food Situation No. 3, October 2012, Global
Information and Early Warning System*, Food and Agriculture
Organization of the United Nations, 2012; www.fao.org/docrep/
014/al980e/al980e00.pdf

'FAO Food Price Index up 1.4 percent in September, Global Cereal
Harvest Down, but Record Expected in LIFDCs', Food and
Agriculture Organization of the United Nations, 4 October 2012;
www.fao.org/news/story/en/item/161602/icode/

'An Agricultural Crime Against Humanity', George Monbiot, *The
Guardian*, 6 November 2007

Time to Eat the Dog?: The Real Guide to Sustainable Living, Brenda
Vale and Robert Vale (Thames & Hudson, 2009), ISBN:
978-0-500-28790-3

'Growing Food in the Desert: Is This the Solution to the World's Food
Crisis?' Jonathan Margolis, *The Observer*, 24 November 2012

'Soil Fertility and Biodiversity in Organic Farming', P. Mäder, A. Fliesbach, D. Dubois, L. Gunst, P. Fried, U. Niggli, *Science*, 2002, 296, pp. 1694–7

One Billion Hungry: Can We Feed the World?, Gordon Conway (Comstock Publishing, 2012), ISBN: 978-0801478024

Darwinian Agriculture: How Understanding Evolution Can Improve Agriculture, R. Ford Denison (Princeton University Press, 2012), ISBN-13: 978-0691139500

Land, Shops and Kitchens: Technology and the Food Chain in Twentieth-century Europe, P. Scholliers, L. Van Molle and C. Sarasua (eds) (Brepols Publishers, 2005), ISBN: 978-2503517803

The End of Food, Paul Roberts (Bloomsbury, 2009), ISBN: 978-0747596424

'The Next 50 Years: Unfolding Trends', Vaclav Smil, *Population and Development Review*, 2005, 31, pp. 605–43

'Magic Beans', Vaclav Smil, *Nature*, 2000, 407, p. 567

'Phosphorus in the Environment: Natural Flows and Human Interferences', Vaclav Smil, *Annual Review of Energy and Environment*, 2000, 25, pp. 53–88

'Nitrogen in Crop Production: An Account of Global Flows', Vaclav Smil, *Global Biogeochemical Cycles*, 1999, 13, pp. 647–62

'Environment: The Disappearing Nutrient', Natasha Gilbert, *Nature*, 2009, 461, pp. 716–18

'Global Food Waste Reduction: Priorities for a World in Transition', J. Parfitt and M. Barthel, Foresight Project on Global Food and Farming Futures, Science Review SR56, UK government Office for Science, 2011; www.bis.gov.uk/assets/foresight/docs/food-and-farming/science/11-588-sr56-global-food-waste-reduction-priorities.pdf

'Food Waste Within Food Supply Chains: Quantification and Potential for Change to 2050', J. Parfitt, M. Barthel, S. Macnaughton, *Philosophical Transactions of the Royal Society B*, 2010, B365, pp. 3065–81

'Save Our Soils', S. Banwart, *Nature*, 2011, 474, pp. 151–2

Reaping the Benefits: Science and the Sustainable Intensification of Global Agriculture (Royal Society, 2009), ISBN: 978-0854037841

'Agricultural Sustainability and Intensive Production Practices', D. Tilman, K.G. Cassman, P.A. Matson, R. Naylor, S. Polasky, *Nature*, 2002, 418, pp. 671–7

www.sundropfarms.com

www.seawatergreenhouse.com

'The Development of C4 Rice: Current Progress and Future Challenges', Susanne von Caemmerer, W. Paul Quick, Robert T. Furbank, *Science*, 2012, 336, pp. 1671–2

Feeding the World: A Challenge for the 21st Century, Vaclav Smil (MIT Press, 2001), ISBN: 978-0262692717

'International Trade in Meat: The Tip of the Pork Chop', James N.
Galloway, Marshall Burke, G. Eric Bradford, Rosamond Naylor,
Walter Falcon, Ashok K. Chapagain, Joanne C. Gaskell, Ellen
McCullough, Harold A. Mooney, Kirsten L.L. Oleson, Henning
Steinfeld, Tom Wassenaar, Vaclav Smil, *AMBIO*, 2007, 36, pp. 622–9
'Phosphorus in the Environment: Natural flows and Human
Interferences', Vaclav Smil, *Annual Review of Energy and the
Environment*, 2000, 25, pp. 53–88
'Global Population and the Nitrogen Cycle', Vaclav Smil, *Scientific
American*, 1997, 277, pp. 76–81
'Harvesting the Biosphere: The Human Impact', Vaclav Smil,
Population and Development Review, 2011, 37 (4), pp. 613–36
'Nitrogen Cycle and World Food Production', Vaclav Smil, *World
Agriculture*, 2011, 2, pp. 9–13
www.incredible-edible-todmorden.co.uk/home
'Changes in Bacterial Community Structure of Agricultural Land Due
to Long-Term Organic and Chemical Amendments', V. Chaudhry,
A. Rehman, A. Mishra, P.S. Chauhan, C.S. Nautiyal, *Microbial
Ecology*, 2012, 64, pp. 450–60
'Ploughing up the Wood-wide Web?', T. Helgason, T.J. Daniell, R.
Husband, A.H. Fitter, J.P.W. Young, *Nature*, 1998, 394, p. 43
'The Environmental Consequences of Adopting Conservation Tillage
in Europe: Reviewing the Evidence', J.M. Holland, *Agriculture,
Ecosystems and Environment*, 2004, 103, pp. 1–25
'Arbuscular Mycorrhizal Fungi as (Agro)ecosystem Engineers', Duncan
D. Cameron, *Plant Soil*, 2010, 333, pp. 1–5
'Benzoxazinoids in Root Exudates of Maize Attract *Pseudomonas
putida* to the Rhizosphere', A.L. Neal, S. Ahmad, R. Gordon-Weeks,
J. Ton, *PLoS ONE*, 2012, 7 (4), e35498; doi:10.1371/journal.
pone.0035498

Chapter 15: Project Sunshine

'The Next 50 Years: Unfolding Trends', Vaclav Smil, *Population and
Development Review*, 2005, 31, pp. 605–43
'Welcome to Dystopia! Entering a Long-term and Politically
Dangerous Food Crisis', Jeremy Grantham, *GMO Quarterly Letter*,
July 2012

Glossary

AC   Alternating current, in which the movement of electric charge periodically reverses direction. Usually found in the mains supply, because it's easier to transport over long distances

Amp   The SI unit of electric current. It's a measure of the amount of electric charge passing a point in an electric circuit per unit of time. One amp is equal to 6.241×10^{18} electrons per second.

AGR   Advanced gas-cooled reactor. A second generation of British gas-cooled reactors, using graphite as a moderator and carbon dioxide as a coolant.

Algal biofuel   An alternative to fossil fuel that uses algae as its renewable source. Like fossil fuel, algal biofuel releases carbon dioxide during combustion, but because this has already been taken out of the atmosphere by the growing algae, the fuel can be considered carbon-neutral.

Alkanes   Saturated hydrocarbon compounds that consist only of hydrogen and carbon atoms and are bonded exclusively by single bonds. Also known as paraffins or kerosene.

Alkenes   Unsaturated hydrocarbon compounds that consist only of hydrogen and carbon atoms but contain carbon atoms joined by both single and double bonds.

AM fungi   Arbuscular mycorrhizal is a type of fungus that penetrates the cortical cells in the roots of vascular plants.

Ammonia   A compound of nitrogen and hydrogen with the chemical formula NH_3. Ammonia is vitally important to life on Earth because of its role in plant nutrition and fertilisation of the soil (either naturally or artificially).

Ammonium nitrate   A compound with the chemical formula NH_4NO_3, commonly used in agriculture as a high-nitrogen fertiliser.

ATP   Adenosine triphosphate transports chemical energy within the cells of living things to fuel their metabolism.

Battery   A device consisting of one or more electrochemical cells that converts stored chemical energy into electrical energy.

Barrel   Unit of volume, as in 'barrels of oil'. One barrel of oil is equal to 159 litres (42 US gallons). The abbreviation is bbl. The extra b in the abbreviation is a throwback to the early days of the oil industry when different containers were used. The 42-US-gallon containers were all blue, so the standard became 'blue barrels of oil' or bbl.

/bbl   Per barrel.

Band gap   The energy range in a solid where no electron states can exist. It's a major factor in determining the electrical conductivity of a solid. Substances with large band gaps tend to be insulators; those with small band gaps are semiconductors; conductors either have tiny band gaps or none at all.

BCE   Before Current Era. An alternative to Before Christ (abbreviated BC).

Biodiesel   A vegetable-oil-based or, occasionally, an animal-fat-based fuel produced by chemically reacting lipids with an alcohol to form fatty acid esters.

Biofuel   A type of fuel whose energy is derived from biological carbon fixation. Includes those derived from biomass and biogases.

Breeder reactor   A breeder reactor is a nuclear reactor capable of generating more fissile material than it consumes because its neutron economy is high enough to breed fissile from fertile material such as uranium-238 or thorium-232.

BRIC   An economic grouping referring to the nations of Brazil, Russia, India and China, which are all considered to be rapidly developing economies.

C3   A metabolic process for carbon fixation in plants via photosynthesis used by most plant species. In C3 plants, carbon dioxide and ribulose bisphosphate are converted into 3-phosphoglycerate. Plants that function using C3 fixation usually thrive in areas where sunlight intensity and temperatures are moderate, ground water is plentiful and concentrations of carbon dioxide are greater than 200 parts per million.

C4   A metabolic process for carbon fixation in plants via photosynthesis used by a small sub-group of plant species (C4 plants) including sugar cane, maize, millet and 46 per cent of all grasses. C4 is more efficient than the more common C3 process and is thought to have evolved more recently. C4 overcomes photo-respiration, which is a wasteful tendency of ribulose bisphosphate to fix oxygen rather than carbon dioxide. C4 plants have a competitive advantage over plants possessing the more common C3 carbon fixation pathway under conditions of drought, high temperatures, and nitrogen or carbon dioxide limitation.

Calorie   Imperial unit of energy, replaced in most fields by the joule, but still widely used when talking about food energy. One calorie = 4.184 J.

Capacitor   A passive electrical component with two terminals, between which energy is stored in an electric field. Unlike batteries, capacitors release their energy very quickly.

Carbohydrate   The most abundant natural organic compound, which consists of carbon, hydrogen and oxygen.

Carbon monoxide   A colourless, odourless and tasteless gas that is slightly lighter than air, with the chemical formula CO. In high concentrations it's toxic to humans and animals. In the atmosphere it's short-lived and a very weak greenhouse gas.

Carbon dioxide   A naturally occurring compound with the chemical formula CO_2. Carbon dioxide is a gas at standard temperature and pressure, and is found as a trace in the atmosphere (in concentrations of 0.039 per cent by volume). Carbon dioxide is a greenhouse gas.

CE   The Current Era. An alternative to Anno Domini (abbreviated AD).

Chelation A molecular process in which two or more separate coordinate bonds between a multiple-bonded ligand and a single central atom are formed.

Coal A combustible fossil fuel occurring as a black or near-black sedimentary rock. Typically found within rock strata in layers or 'seams'. Coal is composed largely of carbon along with varying quantities of other elements, including hydrogen, nitrogen, oxygen and sulphur.

Coal gas A flammable gas used as fuel. It's made by heating (but not burning) coal. Essentially a mixture of hydrogen, carbon monoxide and methane together with small quantities of non-combustible gases such as carbon dioxide and nitrogen.

Conductor A material whose internal electric charges move freely, through which an electric current can be passed.

Coulomb The SI derived unit of electric charge. It's defined as the charge transported by a steady current of 1 ampere in one second.

DC Direct current, in which the movement of electric charge is uni-directional. Direct current is usually produced locally by sources such as batteries, solar cells and dynamos.

Diesel Any liquid fuel used in diesel engines. The most common is a petroleum-derived liquid, but alternatives not derived from petroleum include biodiesel, biomass to liquid (BTL) and biogas to liquid (GTL).

DNA Deoxyribonucleic acid. The information-carrying molecules containing the genetic instructions essential for the development and function of all known living things and most viruses.

EJ Exajoule. Equal to 10^{18} joules.

Electric charge A physical property of matter that causes it to feel a force when near other electrically-charged matter.

Electric current The flow of electrons (electric charge) through a conductor.

Electric potential The difference in electric potential energy of a unit charge transported between two points. Also known as voltage.

Electrical grid An interconnected network for delivering electricity from the point of generation to consumers. There are three components in a grid: power-generation, high-voltage transmission lines to demand centres, and transformers to step down voltage for final delivery.

Electrical resistance The opposition to the passage of an electric current.

Electrolysis A process that uses direct electric current to drive a chemical reaction, such as splitting water molecules into their two constituents: hydrogen and oxygen.

Electromotive force The voltage generated by a battery.

Electrode An electrical conductor used to make contact with a non-metallic part of a circuit (e.g. a semiconductor in a PV cell or a battery).

Electron An elementary particle with a negative electric charge. An electron has a mass that is approximately $1/1836$ of a proton.

Energy, forms of There are two main types of energy: potential, which is a function of the position of an object; and kinetic, which is a function of its motion. In physics, several forms of energy have been identified, all of which are a varying combination of kinetic and potential. They are: thermal (heat), chemical, electric, radiant (electromagnetic radiation), nuclear, magnetic, elastic, acoustic, mechanical, luminous and mass.

Ethanol A volatile liquid with the chemical formula C_2H_6O. Ethanol is a renewable biofuel (usually produced from sugar cane, potato or corn) used mainly as an additive for gasoline. Also known as drinking alcohol.

Exciton A bound state of an electron and hole that are attracted to each other by the electrostatic force.

Fischer–Tropsch process A collection of chemical

reactions that converts a mixture of carbon monoxide and hydrogen into liquid hydrocarbons.

Fertile material   Material that is not itself fissionable by thermal neutrons, but that can be converted into a fissile material by neutron absorption and subsequent nuclei conversions, e.g. thorium-232, uranium-234 and uranium-238.

FFV   Flexible-fuel vehicles are designed to run on more than one fuel, usually gasoline blended with either ethanol or methanol fuel. Both fuels are stored in the same tank.

Fissionable material   A material capable of undergoing fission after capturing a neutron, e.g. uranium-233, uranium-235 and plutonium-239. The group also includes the subset 'fissile material', which is any material capable of sustaining a chain reaction of nuclear fission.

Force   Any influence that causes an object to undergo change, either in motion, direction or geometrical construction.

Forces, fundamental   The four fundamental forces of nature are: the strong nuclear force, electromagnetism, the weak nuclear force, and gravity.

The strong nuclear force binds together protons and neutrons in the nucleus of an atom. It's also the force that holds quarks together to form protons, neutrons and other hadron particles. It's the strongest – about 100 times greater than electromagnetism.

Electromagnetism is the forces that occur between electrically charged particles.

The weak nuclear force is responsible for the radioactive decay of sub-atomic particles (beta decay) and initiates the process known as stellar nucleosynthesis in stars. It's several orders of magnitude weaker than electromagnetism.

Gravity is the force by which physical objects attract each other that is proportional to their masses. It's by far the weakest of all.

Fossil fuels Fuels formed by the anaerobic decomposition of long-dead organisms.

Fracking Hydraulic fracturing. A process that involves pumping high-pressure water into fractures in bedrock to liberate stores of natural gas.

GM Genetic modification.

Greenhouse gas (abbreviated GHG) A gas in the atmosphere that absorbs and emits radiation within the thermal infrared range. This process is the fundamental cause of the greenhouse effect.

GW Gigawatt or 1 billion watts.

GWh Gigawatt hour.

Haber–Bosch process An industrial procedure for producing ammonia through a reaction of nitrogen and hydrogen.

Hertz The SI unit of frequency, defined as the number of cycles per second.

HEV Hybrid electric vehicle, which combines a conventional internal combustion engine with an electric motor system to achieve better fuel economy than a conventional vehicle.

Hydrogen Containing one proton and one electron, hydrogen is the simplest element with the symbol H and atomic number 1. It's the most abundant chemical substance in the universe, accounting for around 75 per cent of all matter. Hydrogen has two isotopes: deuterium is stable and contains one proton and one neutron in its nucleus; tritium contains one proton and two neutrons in its nucleus and is slightly radioactive.

Insolation A measure of solar radiation energy received on a given surface area and recorded during a given time.

Insulator A material whose internal electric charges do not flow freely and therefore does not conduct electric current.

Intermittency An intermittent energy source is one that is not continuously available due to some factor outside direct control. Intermittency is a problem that affects most sources of renewable energy

Joule   The SI unit of energy, work or heat. Defined as the work done by applying a force of 1 newton through a distance of 1 metre, or the work required to move an electric charge of 1 coulomb through an electrical potential difference of 1 volt.

J/g   Joules per gram.

kJ   Kilojoule or 1,000 joules.

kJ/s   Kilojoules per second.

kW   Kilowatt or 1,000 watts.

kW/m²   Kilowatts per square metre.

kWh   Kilowatt hour.

Li-air   A lithium-air battery uses the oxidation of lithium at the anode and reduction of oxygen at the cathode to induce a current.

Li-ion   Lithium-ion batteries are a type of rechargeable battery in which lithium ions travel from the negative to positive electrodes during discharge, and vice versa when charging.

LFTR   A liquid fluoride thorium reactor is a thermal breeder reactor that uses the thorium fuel cycle. Graphite rods function as a moderator and molten salts provide both a stable coolant and a fuel supply. The system is based on designs for a molten salt reactor (MSR).

LPG   Liquefied petroleum gas, also called propane or butane. It's a mixture of flammable hydrocarbon gases used as a liquid fuel.

LWR   A light water reactor is a type of thermal nuclear reactor that uses ordinary water as both a coolant and neutron moderator. Thermal reactors are the most common type of nuclear reactor, and light water reactors are the most common type of thermal reactor. There are three types of light water reactor: the pressurised water reactor (PWR), the boiling water reactor (BWR), and the supercritical water reactor (SCWR).

Methane   The main component of natural gas with the chemical formula CH_4. It's the simplest alkane and one of the most abundant organic compounds on the planet. Methane is also a greenhouse gas.

Methanol   The simplest form of alcohol, a volatile liquid with the chemical formula CH_4O. Methanol is an alternative liquid fuel and can be used either as an additive for gasoline or on its own. Methanol is more toxic than ethanol (although not much more than gasoline) but is less expensive to produce sustainably.

Mg-Sb battery   A magnesium-antimony battery is a form of high-temperature electric battery that uses molten salts as an electrolyte.

MJ   Megajoule or 1,000,000 joules.

MSR   A molten salt reactor is a class of nuclear fission reactor in which the primary coolant and the fuel itself is a molten salt mixture. MSRs operate at a higher temperature than water-cooled reactors and are more efficient.

MW   Megawatt or 1,000,000 watts.

Natural gas   A combustible fossil fuel occurring as a hydrocarbon gas mixture of methane (primarily) along with other hydrocarbons, carbon dioxide, nitrogen and hydrogen sulphide.

Nuclear fission   A process in which the nucleus of an atom is split into smaller parts, thereby releasing a huge amount of energy.

Nuclear fusion   A reaction in which two or more atomic nuclei fuse together to form a single heavier nucleus, releasing enormous amounts of energy locked away by the strong nuclear force.

Neutron   A sub-atomic particle that has no electric charge and a mass slightly larger than that of a proton. One or more neutrons are present in the nucleus of each atom (except for hydrogen).

Ohm   The SI unit of electrical resistance. Defined as a resistance between two points of a conductor when a constant potential difference of 1 volt applied to these points produces in the conductor a current of 1 amp.

OPV   Photovoltaic cell using organic semiconductors.

Oxidation   The loss of electrons or an increase in oxidation state by a molecule, atom or ion.

Petroleum   A combustible fossil fuel, occurring as a flammable liquid consisting of a complex mixture of hydrocarbons of various molecular weights and other liquid organic compounds that are found in geological formations beneath the Earth's surface.

Photodetector   Material sensitive to light or other electromagnetic energy.

Photoelectric effect   The process by which the absorption of energy from electromagnetic radiation causes electrons to be emitted from matter.

Photon   An elementary particle, the quantum of light and all other forms of electromagnetic radiation, and the force-carrier for the electromagnetic force, even when static via virtual photons. Functions as a particle and a wave.

Photosynthesis   The process used by plants (and some other organisms) to convert sunshine (photons) into chemical energy that can be stored as fuel for later use by the organism.

Photovoltaic effect   The creation of voltage or electric current in a material upon exposure to light.

Plasma   A distinct phase of matter: not a solid, liquid or a traditional uncharged gas, it comprises charged particles and free electrons that respond strongly and collectively to electromagnetic fields.

Plutonium   A transuranic radioactive chemical element with the symbol Pu and atomic number 94. Two isotopes, plutonium-239 and plutonium-241, are both fissile and can sustain a nuclear chain reaction.

p-n junction   In a single crystal of semiconductor, a p–n junction is formed by intentionally introducing impurities into the material to create an active site within which electronic activity can take place. P-n junctions are fundamental components of most semiconductor devices such as diodes, LEDs, transistors and PV cells.

Potassium nitrate   A salt with the chemical formula KNO_3. Known as saltpetre, it's used as a constituent of fertilisers.

Proton   A sub-atomic particle with a positive electric charge. One or more protons are present in the nucleus of each atom.

PV cell   Abbreviation of photovoltaic cell, a solar-cell that converts light energy directly into electricity via the photovoltaic effect.

Quantum dot   Solar cells employing low band gap semiconductor crystals that are so small they form quantum dots instead of organic dyes as light absorbers.

Reduction   The gain of electrons or a decrease in oxidation state by a molecule, atom or ion.

RNA   Ribonucleic acid performs multiple vital roles in the coding, decoding, regulation and expression of genes. The medium through which genetic information is transmitted.

Semiconductor   A material whose electric conductivity is between that of a conductor and an insulator. Electric current conduction in a semiconductor occurs via free electrons and the positively-charged holes they leave, known collectively as charge-carriers.

Shockley–Queisser limit   The maximum theoretical efficiency of a solar cell using a semiconductor junction to collect power from the cell.

Silicon   A chemical element with the symbol Si and atomic number 14. In monocrystalline form (single-crystal Si or mono-Si) it's the base material of the electronics and computer industries.

Sodium nitrate   A salt with the chemical formula $NaNO_3$. Sodium nitrate is also known as 'Chilean saltpetre' and is used as a constituent of fertilisers.

Syngas   A gas mixture that contains varying amounts of carbon monoxide and hydrogen.

Thorium   Thorium is a naturally occurring weakly radioactive chemical element with the symbol Th and atomic number 90. Its natural isotope thorium-232 can be used as fuel in nuclear reactors and breeder reactors.

Tonne   The SI unit of mass equal to 1,000 kilos (2,204.6 lb). Not to be confused with either the Imperial ton (2,240 lb) or US ton (2,000 lb).

TW   Terawatt or 1 trillion watts.

TWh   Terawatt hour.

Uranium   Uranium is a weakly radioactive metal with the symbol U and the atomic number 92. All of its isotopes are unstable. It occurs naturally as U-238 (99.27 per cent) and U-235 (0.72 per cent) with a very small amount of U-234 making up the remainder. Uranium decays slowly by emitting alpha particles.

Volt   The SI unit of electrical potential, electrical potential difference and electromotive force. Defined as the difference in electric potential across a wire when an electric current of 1 ampere dissipates 1 watt of power.

Watt   The SI unit of power that measures the rate of energy or transfer. One watt is equal to 1 joule per second.

Work   When a force acts on a body so that there is a displacement of the point of application, however small, in the direction of the force, it's said to be doing work.

Index